安徽省高等学校"十三五"省级规划教材
科技英语丛书

采矿工程专业英语

第 2 版

English for Coal Mining Engineering

主　编　杨　科
副主编　刘钦节
参　编（按姓氏拼音顺序）
　　　　陈登红　李　宁　李传明　李英明
　　　　秦汝祥　殷志强　张　通　张向阳

中国科学技术大学出版社

内 容 简 介

本书内容包含煤矿地质与测量、矿山岩石力学、矿井开拓、井巷施工与支护、采煤工艺与采煤方法、矿山压力及其控制、矿井通风、矿井辅助生产系统、开采沉陷与控制、煤与瓦斯共采、露天开采、洁净煤技术与环境保护、矿山安全与职业健康、煤炭智能精准开采、煤矿固体废弃物综合利用、未来采矿技术等，反映了当前煤炭工业的新技术和新发展。每个单元皆有常用表达或一些短语的注解。本书还介绍了科技英语的特点，较为系统和详细地讲解了科技英语写作技巧，简单介绍了科技论文中英文摘要和结论的写作要点；列举了国内外采矿行业、大学、期刊等学术信息；有利于学生进一步理解和掌握。

本书无论是在内容选材还是在编写上都具有专业特色和学术价值，是一本实用的高等院校采矿专业本科教材，也可供采矿工程专业研究生及从事采矿工程专业的技术或管理人员学习参考。

图书在版编目（CIP）数据

采矿工程专业英语/杨科主编. —2 版. —合肥：中国科学技术大学出版社，2020.12
ISBN 978-7-312-05121-0

Ⅰ. 采… Ⅱ. 杨… Ⅲ. 矿山开采—英语—高等学校—教材 Ⅳ. TD8

中国版本图书馆 CIP 数据核字(2020)第 253352 号

采矿工程专业英语
CAIKUANG GONGCHENG ZHUANYE YINGYU

出版	中国科学技术大学出版社
	安徽省合肥市金寨路 96 号，230026
	http://press.ustc.edu.cn
	https://zgkxjsdxcbs.tmall.com
印刷	安徽国文彩印有限公司
发行	中国科学技术大学出版社
经销	全国新华书店
开本	710 mm×1000 mm　1/16
印张	13.75
字数	360 千
版次	2012 年 8 月第 1 版　2020 年 12 月第 2 版
印次	2020 年 12 月第 2 次印刷
定价	39.00 元

Preface 前言

《采矿工程专业英语》自出版以来，受到很多煤炭类高校采矿工程专业师生的青睐，并收获了很多好评。然而，随着新一轮科技革命和产业变革加速进行，第四次工业革命正以指数级速度展开，全球主要经济体正在寻求发展新契机，综合国力竞争愈加激烈。我国经济发展进入新常态，为在未来全球创新生态系统中占据战略制高点，我国密集开展实施了"中国制造2025""互联网＋""网络强国"等创新驱动发展重大战略。针对采矿工程这种传统的工科专业，各相关高校皆从自身定位与特色出发，面向工业界、面向世界、面向未来的实际需要，主动通过信息化、智能化或其他学科的渗透而转型、改造和升级构建产出导向的教育（Outcomes-Based Education，简称"OBE"）新模式，以有效应对新一轮科技革命和产业变革挑战。

为此，本书编者团队在深入分析新工科、基于OBE理念的采矿工程专业人才培养目标与方案的基础上，明确了新时代背景下专业英语对采矿工程专业大学生培养目标达成的支撑情况，构建了基于OBE理念的采矿工程专业英语教材建设模型，明确了专业英语教材建设的结构体系，进而对第1版进行了修改和完善，以期能够更好地为采矿工程专业大学生相关课程教与学提供重要参考。

与第1版相比，第2版具有如下几个特点：

1. 本书共有16个单元和2个附录，主要内容包括煤矿地质与测量、矿山岩石力学、矿井开拓、井巷施工与支护、采煤工艺与采煤方法、矿山压力及其控制、矿井通风、矿井辅助生产系统、开采沉陷与控制、煤与瓦斯共采、露天开采、洁净煤技术与环境保护、矿山安全与职业健康、煤炭智能精准开采、煤矿固体废弃物综合利用、未来采矿技术等内容，更充分地反映了当代专业英语课程建设与相关行业发展的最新成果，可为培养适应我国社会快速发展的采矿专业国际化人才提供帮助。

2. 本书不仅强化了煤炭资源开采的基础知识，而且增加了近5年煤炭资源开采的新理论、新方法、新技术，特别是数字矿山、安全智能精准采煤等学科前沿知识。对于国外成熟技术与方法均通过选择、改编、整合、补充、扩展等方式，对国家规划教材、采矿工程学科进展、国内外最新出版的专业文章等相关内容进行再加工、再创造。

3. 每个单元皆设有常用表达，并将煤炭工业相关专业知识、国内最新技术进展、专业英语词汇及表达方式、科技英语写作等内容有机地融为一体，丰富学生知识面,扩展学生视野。

　　本书由安徽理工大学矿业工程学院的编者团队进行修订和完善,杨科教授任主编,刘钦节副教授任副主编,其他参编者还有张向阳副教授、殷志强副教授、李英明教授、李传明副教授、秦汝祥副教授、陈登红副教授、李宁博士、张通博士。本书具体分工如下:杨科负责前言、概述、第 14 单元、附录 A;刘钦节负责第 1、15、16 单元;张向阳负责第 5、6 单元;殷志强负责第 2 单元;李英明负责第 3、4 单元;李传明负责第 9、11 单元;秦汝祥负责第 7、13 单元;陈登红负责第 8 单元;李宁负责第 10 单元、附录 B;张通负责第 12 单元。

　　在新书出版之际,对本书编写及修订过程中参考的国内外文章、书籍等资料的作者表示衷心的感谢!感谢美国国家工程院院士、西弗吉尼亚大学采矿工程系资深教授 Syd S. Peng 博士和罗毅教授为本书提供了素材和帮助。感谢付强、魏祯及第 1 版教材出版过程中在文字录入、绘图、排版等方面做出辛苦工作的研究生。感谢国家级采矿工程专业综合改革试点、教育部"新工科"研究与实践项目"新工科背景下采矿工程传统优势专业改造升级探索与实践"、安徽理工大学一流本科人才示范引领基地、安徽省《采矿工程专业英语》省级规划教材等质量工程项目以及安徽省煤炭安全智能精准开采教学团队对本书出版的经费支持!

　　由于时间仓促,加之本书内容新、涉及面广以及编者水平有限,错漏、不足之处在所难免,恳请广大读者及相关专家批评指正。

编　者

2020 年 8 月

Contents 目录

Preface 前言 ·· (ⅰ)

Introduction to Coal Mining 煤炭开采概述 ······························· (1)

Unit 1　Coal Mine Geology 煤矿地质 ······································ (5)
　　Text 1　Geologic Structure of Coalfield　煤田地质构造 ········· (6)
　　Text 2　The Occurrence of Coal Seam　煤层产状 ················ (10)
　　Terms　常见表达 ··· (11)

Unit 2　Rock Mechanics for Mining 矿山岩石力学 ···················· (14)
　　Text 1　Rock Mechanics for Mining　矿山岩石力学 ············· (14)
　　Text 2　Uniaxial Compression Test of Isotropic Rock　各向同性岩石
　　　　　　单向抗压测试 ··· (17)
　　Terms　常见表达 ··· (19)

Unit 3　Mine Development 矿井开拓 ······································ (22)
　　Text 1　Mine Field Development　矿区开发 ······················· (23)
　　Text 2　Portal Design　井筒设计 ····································· (25)
　　Terms　常见表达 ··· (29)

Unit 4　Shaft Sinking, Drifting and Roadway Support 井巷施工与支护 ··· (32)
　　Text 1　Shaft Sinking Technology　井筒开凿技术 ················ (32)
　　Text 2　Drifting Technology　掘进技术 ····························· (35)
　　Text 3　Roadway Support　巷道支护 ································ (38)
　　Terms　常见表达 ··· (40)

Unit 5　Coal Mining Technology and Method 采煤工艺与采煤方法 ········· (44)
　　Text 1　Classification of Mining Operation　采矿作业的分类 ······· (44)
　　Text 2　Coal Mining Methods　采煤方法 ··························· (45)
　　Text 3　Coal Mining Technology in Longwall Mining　长壁采煤法 ······ (47)

Terms　常见表达 …………………………………………………………（50）

Unit 6　Rock Pressure and Strata Control　矿山压力及其控制 …………（54）
Text 1　Rock Pressure in Mine　矿压显现 ………………………………（54）
Text 2　Strata Control in Underground Mining　矿压控制 ……………（58）
Text 3　Supports in Longwall Mining　长壁开采支架 …………………（62）
Terms　常见表达 …………………………………………………………（65）

Unit 7　Mine Ventilation　矿井通风 ……………………………………（68）
Text 1　Imfortance of Mine Ventilation Systems　矿井通风系统的必要性 …………………………………………………………（68）
Text 2　Mine Ventilation Systems　矿井通风系统 ……………………（72）
Terms　常见表达 …………………………………………………………（78）

Unit 8　Mine Auxiliary Production System　矿井辅助生产系统 ………（80）
Text 1　Mine Haulage System　矿井运输系统 …………………………（81）
Text 2　Mine Hoisting System　矿井提升系统 …………………………（82）
Text 3　Mine Drainage System　矿井排水系统 …………………………（84）
Terms　常见表达 …………………………………………………………（86）

Unit 9　Mine Subsidence and Control　开采沉陷与控制 ………………（88）
Text 1　Introduction to Mining Subsidence　开采沉陷概述 …………（88）
Text 2　Mining Subsidence and Ground Control Technology　开采沉陷与矿压控制技术 ……………………………………………（91）
Terms　常见表达 …………………………………………………………（95）

Unit 10　Integrated Coal Mining and Methane Extraction　煤与瓦斯共采 …………………………………………………………………（96）
Text 1　Integrated Coal Mining and Methane Extraction　煤与瓦斯共采 …………………………………………………………（96）
Text 2　Integrated Coal Mining and Methane Extraction (Continued)　煤与瓦斯共采(续) ……………………………………………（100）
Terms　常见表达 …………………………………………………………（105）

Unit 11　Surface Mining　露天开采 ……………………………………（106）
Text 1　Introduction to Surface Mining　露天开采概论 ………………（106）
Text 2　Surface Coal Mining Methods in China　中国的露天煤矿开采方法 ……………………………………………………………（111）

Terms 常见表达 ………………………………………………… (113)

Unit 12 Clean Coal Technology and Environment Protection 洁净煤技术与环境保护 ………………………………………… (116)
Text 1 Coal Mine and the Environment 矿山与环境 ……………… (116)
Text 2 Clean Coal Technologies 洁净煤技术 …………………… (119)
Text 3 Plan of Clean Coal Utilization 洁净煤利用规划 ………… (123)
Terms 常见表达 ………………………………………………… (124)

Unit 13 Mine Safety and Occupational Health 矿山安全与职业健康 …… (127)
Text 1 Mine Safety 矿山安全 ……………………………………… (127)
Text 2 Introduction to Occupational Health and Safety 职业健康与安全概述 ……………………………………………… (132)
Terms 常见表达 ………………………………………………… (136)

Unit 14 Intelligent Precise Coal Mining 煤炭智能精准开采 …………… (139)
Text 1 Intelligent and Ecological Coal Mining Technology 智能生态采矿技术 ……………………………………………… (140)
Text 2 Future of China Coal Industry 中国煤炭工业的未来 ……… (144)

Unit 15 Comprehensive Utilization of Mining and Mineral Processing Wastes 煤矿固体废弃物综合利用 ………………………… (147)
Text 1 Utilization Techniques of Mining and Mineral Processing Wastes 煤矿固体废弃物利用技术 ……………………… (148)
Text 2 Solid Backfill Mining Techniques 固废充填采矿技术 ……… (152)

Unit 16 Future Mining Technology 未来采矿技术 ………………………… (155)
Text 1 A Strategic Approach for Sustainable Mining in Future 煤矿未来可持续发展策略 ………………………………… (155)
Text 2 Digital Mine 数字化矿山 …………………………………… (159)
Text 3 Automation Mining Technology 无人采矿技术 …………… (162)
Text 4 Virtual Reality (VR) Mining 虚拟采矿技术 ……………… (166)
Terms 常见表达 ………………………………………………… (169)

Appendix A English Writing of Science and Technology 科技英语写作 …………………………………………………………… (171)
A.1 Introduction 概论 ……………………………………………… (171)
A.2 Standards and Skills of Language 语言表达的规范与技巧 …… (176)

A. 3　Contents of Science and Technology Paper　英语科技论文内容 ……………………………………………………………………（183）

A. 4　Abstract Writing　摘要的撰写 ……………………………（198）

Appendix B　Academic Information for Mining　采矿专业学术信息………（205）

References　参考文献 …………………………………………………（209）

Introduction to Coal Mining

煤炭开采概述

Coal is one of the world's most important resources of energy, accounting for the largest share of power generation at 38%, the same share as 20 years ago. Coal is a fossil fuel and is far more plentiful than oil or gas, with coal reserves worldwide in 2018 accounting for 132 years of production. Not only does coal provide electricity, it is also an essential fuel for steel and cement production, and other industrial activities.

1. What is Coal?

Coal is a fossil fuel, which is a combustible, sedimentary, organic rock, and it is composed mainly of carbon, hydrogen and oxygen. It is formed from vegetation, which has been consolidated between other rock strata and altered by the combined effects of pressure and heat over millions of years to form coal seams.

The build-up of silt and other sediments together with movements in the earth's crust (known as tectonic movements) buried these swamps and peat bogs, often to great depths. With burial, the plant material was subjected to high temperatures and pressures. This caused physical and chemical changes in the vegetation, transforming it into peat and then into coal. Coal formation began during the Carboniferous Period (known as the first coal age), which spanned 360 million to 290 million years ago.

The quality of each coal deposit is determined by temperature and pressure and by the length of time in formation, which is referred to as its "organic maturity". Initially the peat is converted into lignite or "brown coal" — these are coal-types with low organic maturity. In comparison to other coals, lignite is quite soft and its color can range from dark black to various shades of brown.

Over many more millions of years, the continuing effects of temperature and pressure produces further change in the lignite, progressively increasing its organic maturity and transforming it into the range known as "sub-bituminous" coals.

Further chemical and physical changes occur until these coals became harder and blacker, forming the "bituminous" or "hard coals". Under the right conditions, the progressive increase in the organic maturity can continue, finally forming anthracite.

2. Importance of Coal

Due to its abundance, coal has been mined in various parts of the world throughout history and continues to be an important economic activity today. Compared to wood fuels, coal yields a higher amount of energy per mass and could be obtained in areas where wood is not readily available. Though historically used as a means of household heating, coal is now mostly used in industry, especially in smelting and alloy production, as well as electricity generation.

Coal plays a vital role in meeting global energy needs and is critical to infrastructure development — 38% of the world's electricity, the same share as 20 years ago, and 71% of the world's steel is produced with coal as its fuel. Since 2000, coal production in China has increased by 161.8%. In 2018, 7813.3 million tons (Mt) of coal were produced worldwide, primarily mined by two methods: surface or "opencast" mining, and underground mining. The largest coal producing countries are not confined to one region — the top five hard coal producers are China, India, USA, Indonesia and Australia. Much of global coal production is used in the country in which it was produced; only around 15% of hard coal production is destined for the international coal market.

Table 1 Top ten hard coal producers in 2018

Producer	Coal production	Producer	Coal production
China	3550 Mt	Russia	420 Mt
India	771 Mt	South Africa	259 Mt
USA	685 Mt	Germany	169 Mt
Indonesia	549 Mt	Poland	122 Mt
Australia	483 Mt	Kazakhstan	114 Mt

(Source: World Coal Association 2018, https://www.worldcoal.org/coal/coal-mining)

3. History of Coal Mining

The Industrial Revolution, which began in Britain in the 18th century, and later spread to continental Europe and North America, was based on the availability of coal to power steam engines. International trade expanded exponentially when coal-fed steam engines were built for the railways and steamships. The new mines that grew up in the 19th century depended on men and children to work long hours in often dangerous working conditions. There were many coalfields, but the oldest were in Newcastle and Durham, South Wales, Scotland and the Midlands, such as those at Coalbrookdale.

The oldest continuously worked deep-mine in the United Kingdom is Tower Colliery in South Wales valleys in the heart of the South Wales coalfield. This colliery was developed in 1805, and its miners bought it out at the end of the 20th century, to prevent it from being closed. Tower Colliery was finally closed on 25 January 2008, although production continues at the Aberpergwm drift mine owned by Walter Energy of the USA nearby.

Coal was mined in America in the early 18th century, and commercial mining started around 1730 in Midlothian, Virginia. Mechanization of operations at the face started before 1900 with development of punching machines and chain-type cutters for undermining the coal seam before blasting, of coal and rock drills, electric and compressed air locomotives, and even some early experiments with continuous mining machines.

Longwall mining was used here and there in the USA until about 1910, particularly in Illinois, but then became noncompetitive with room-and-pillar methods in thicker seams that better lent themselves to mechanization. In the meantime, longwall continued to be dominant in Europe and Asia because of thin coal and depth of cover.

Coal-cutting machines were invented in the 1880s. Before this invention, coal was mined from underground with apick and shovel. By 1912 surface mining was conducted with steam shovels designed for coal mining.

4. Problems in Coal Mining

Despite the tremendous importance of coal, the industry is faced with serious problems, such as:

(1) Dust and Noise Pollution

(i) Dust at mining operations can be caused by trucks driven on unsealed roads, coal crushing operations, drilling operations and wind blowing over areas disturbed by mining.

(ii) Main sources of noise pollution are blasting, movement of heavy earth moving machines, drilling and coal handling plants.

(2) Mining Subsidence

Mine subsidence can be a problem with underground coal mining, wherein the ground level lowers as a result of coal having been mined beneath. The major factors affecting the extent of subsidence are seam thickness and its depth beneath the surface. Roof collapse will occur within 24 hours of coal extraction, but the full effects are transmitted rather slowly upwards, eventually resulting in subsidence at the surface. But it may take over 10 years before the surface is completely stable again.

(3) Water Pollution

Most underground and some surface mines lie well below the water table. Both, therefore, have the potential to pollute any groundwater that flows through them. Waste water from coal preparation plant and mine water are other sources of water pollution.

In addition to the obvious disturbance of the land surface, mining may affect, to varying degrees, groundwater, surface water, aquatic and terrestrial vegetation, wildlife, soils, air, and cultural resources. Actions based on environmental regulations may avoid, limit, control, or offset many of these potential impacts, but mining will, to some degree, always alter landscapes and environmental resources. Regulations intended to control and manage these alterations of the landscape in an acceptable way are in place and continually updated as new technologies are developed to improve mineral extraction, to reclaim mined lands, and to limit its impact on environment.

Unit 1　Coal Mine Geology

煤 矿 地 质

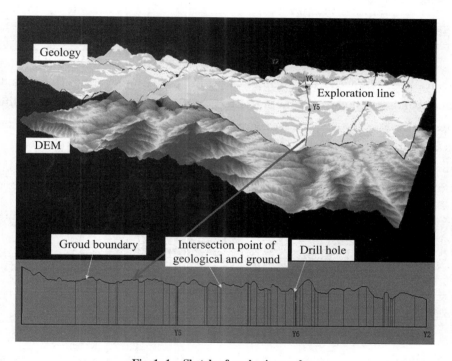

Fig. 1.1　Sketch of coal mine geology

Coal is a combustible sedimentary rock formed from ancient vegetation which has been consolidated between other rock strats and transformed by the combined effects of microbial action, pressure and heat over a considerable time period. This process is commonly called "coalification". Coal occurs as layers or seams, ranging in thickness from millimeters to many tens of meters. Coal reserves are discovered through exploration. Modern coal exploration typically involves extensive use of geophysical surveys, including 3D seismic surveys aimed at providing detailed information on the structures with the potential to affect longwall operations, and drilling to determine coal quality and thickness.

Text 1 Geologic Structure of Coalfield 煤田地质构造

Many geologic factors influence thickness, continuity, quality, and mining conditions of coal. Some features formed during peat accumulation or shortly thereafter, whereas others developed long after deep burial and coalification. Several more common features that have been observed are described in the following sections:

1. Shale Partings in Coal and Split Coal

In the peat swamp in which vegetable material accumulated, streams periodically overflowed their banks, depositing thin mud and silt layers that became bands of shale or siltstone after the vegetable material was coalified. Particularly close to the flooding rivers, relatively thick deposits were laid down, with reestablishment of the swamp vegetation after the flood stage, and this resulted in one type of split coal observed in coalfields (Fig. 1. 2).

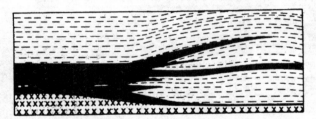

Fig. 1. 2 Splitting of a coal seam

2. Washouts

After plant material accumulates and is buried by various sediments, it may be removed by the downward erosive action of streams. Its absence in the rock sequence is termed a washout or cutout (Fig. 1. 3). Washouts may happen shortly after deposition of the peat or much later, even after the coal has been covered by other sediments. The channel is later filled with mud and sand deposits so that the normal position of the coal is occupied by shale or sandstone.

3. Folds

Throughout geologic time, pressed by horizontal forces, coal and rock strata distort into a continuous wavy forms, this structure is known as a fold (Fig. 1.4).

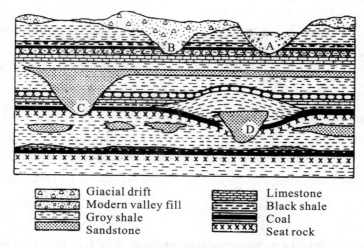

Fig. 1.3 Some features affecting continuity of coals

Coal removed by modern stream erosion at A; pre-glacial erosion at B; by a stream after coal deposition at C; and at D — the stream was present throughout the time of peat accumulation.

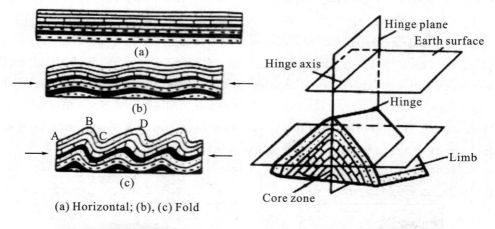

(a) Horizontal; (b), (c) Fold

Fig. 1.4 Folding and folds   Fig. 1.5 The parameters of fold

The basic unit of a fold is named a fold unit which is one bending. Fold units have two patterns: closing upwards in the plane of the fold profile are known as termed anticlines, such as A, B, C in Fig. 1.4; synclines folds are those closing downwards in the same plane, such as B, C, D in Fig. 1.4.

The fold unit is described by fold parameters illustrated in Fig. 1.5. The parameters include: core zone, limb zone, hinge surface, hinge axis, fold hinge, etc.

Fold limb: beside the hinge zone.

Hinge surface: a surface dividing the two fold limbs in the middle, and the surface may be vertical, inclined, or horizontal.

Hinge axis: the intersection line between hinge surface and the earth's surface, hinge axis may be a straight or curved line, the direction of the hinge axis indicates the direction of the fold extension.

Fold hinge: the intersection line between a rock stratum surface and the hinge surface, and it may be horizontal, inclined, or wavy. Fold hinge can indicate the fold occurrence changes along the direction of the fold extension.

4. Faults

Faults are fractures in the rock sequence along which strata on each side of the fracture appear to have moved in different directions. The movement may be measured in inches, tens or hundreds of feet, or less commonly, in miles. Movement may be in any direction, from horizontal to vertical. The two most common types of faults observed are illustrated in block diagrams shown in Fig. 1.6. Where stresses are in opposite directions or tensional, rocks have "pulled apart" at the fracture surface and displacement is as illustrated for a normal fault [Fig. 1.6(a)]. Where horizontal compressive forces are responsible for faulting, one block may be showed over the other in the manner illustrated in Fig. 1.6(b) to produce a thrust or reverse fault. A borehole drilled very close to a normal fault may have a shortened interval between key strata, whereas in a reverse fault, part of the sequence may be repeated in the boring.

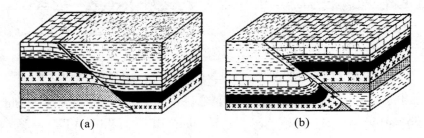

Fig. 1.6　Faults

With normal fault (a) strata above fault have moved down relative to those above; with reverse fault (b) strata above have moved up.

The faults are divided into three kinds according to the relationship between the fault planes strike direction and rock strike direction (Fig. 1.7). If the fault plane strike is paralleled to the rock strike, the fault is known as a strike fault. If the fault plane strike is perpendicular to the rock strike, the fault is known as a dip fault. If the fault plane strike is intersected with a rock strike, the fault is known as an intersected fault.

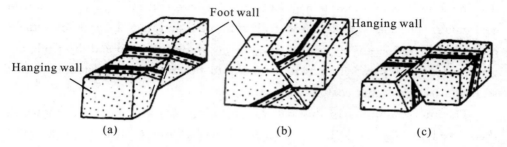

Fig. 1.7 Block diagram illustrating the three types of faults
(a) Normal dip-slip fault; (b) Reverse dip-slip fault; (c) Strike-slip fault

5. Clastic Dikes or Clay Veins

Irregular, vertical-to-inclined tabular masses of clastic material (clay, silt, or sand) that interrupt the coal seam are called clastic dikes. They have also been called horsebacks, a name also applied in some areas to rolls. They may be from a fraction of an inch to several feet wide and usually extend laterally for many feet to hundreds of feet. They may extend for some distance into the strata overlying the coal, and frequently contribute to a weakened roof condition when the coal is mined.

6. Cleat and Fractures

Bituminous coals commonly exhibit numerous closely spaced, relatively smooth vertical fractures, or cleat, spaced a fraction of an inch to a few inches apart. Very commonly, there are two preferred directions to the trend of the cleat, at approximately 1.57 rad (90°) to each other. The cleat that is better developed is termed the face cleat, and the other is the butt cleat.

7. Concretion

The coal as well as the associated rocks commonly contains aggregations of minerals in spherical, disklike, or irregular forms. They may be microscopic or several feet across, although the most commonly observed size is several inches wide.

Text 2　The Occurrence of Coal Seam　煤层产状

Determining the occurrence of coal seams, referring to the spatial distribution of coal seams, is a prerequisite for safe and efficient production in coal mines. The coal seam is flat when it is formed, and then the crust movement causes many coal seams to incline. In general, the strike, dip and dip angle are the main deposit parameters that are used to describe the occurrence of coal seams, as well as other coal measure strata.

The intersection line of the coal seam plane and horizontal plane is known as the strike line (Fig. 1.8). The direction of the strike line is called the strike. The strike is described by its directional bearing, such as west-east strike, north-south strike, southeast-northwest strike, etc.

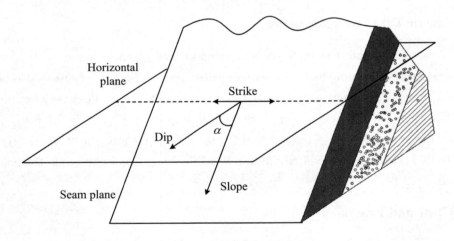

Fig. 1.8　Schematic diagram of coal seam occurrence

The line which intersects the strike line at a right angle in the coal seam plane is known as the dip line, the direction of the dip line, from up to down is called the dip. The projection of the dip on the horizontal plane is called the dip direction (Fig. 1.8), which is described by its directional bearing, such as south dip, north dip, south-west dip, etc.

The intersecting angle between the coal seam plane and the horizontal plane is known as the dip angle, symbol a represents the dip angle in Fig. 1.8. The dip angle of coal seam varies between $0° \sim 90°$.

Based on mining technology, coal seams are divided into three kinds by three dip angles:

Gently inclined coal seam: dip angle 0°~25°.

Inclined coal seam: dip angle 25°~45°.

Steeply pitching coal seam: dip angle 45°~90°.

The coal seam floor contour map is the most important geological map in the production of coal-mine, which is the base of design, production and reserve-counting of coal-mine, as is shown in Fig. 1.9. Therefore, it is often necessary to determine the occurrence of the coal seam through the contour map of coal seam floor.

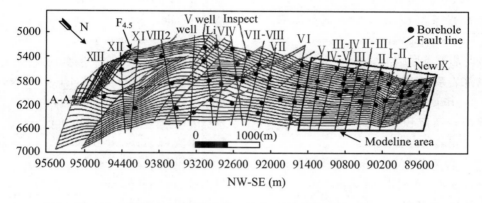

Fig. 1.9 The coal seam floor contour map

It can be determined from Fig. 1.9 that, the northern part of the coalmine is a monoclinic structure with the strike direction along northwest and the dip direction toward northeast. The southern part of the coal mine is a recumbent structure where the folds are overturned and the stratigraphic sequence of upper limbs is reversed. The dips of the strata and the coal seams are approximately 15°~25°.

Terms 常见表达

煤层 coal seam, coal bed
煤层厚度 thickness of coal seam
最低可采厚度 minimum minable thickness
有益厚度 profitable thickness
煤层结构 texture of coal seam
可采煤层 minable coal seam, workable coal seam

煤层形态 form of coal seam, occurrence of coal seam
煤层分叉 splitting of coal seam, bifurcation of coal seam
煤层尖灭 thin-out of coal seam, pinch-out of coal seam
煤相 coal facies

煤核 coal ball
夹矸 parting, dirt band
煤层冲刷 washout
煤组 coal seam group
煤沉积模式 sedimentary model of coal
含煤岩系 coal-bearing series, coal measures
近海型含煤岩系 paralic coal-bearing series
内陆型含煤岩 inland coal-bearing series
浅海型含煤岩系 neritic coal-bearing series
含煤岩系成因标志 genetic marking of coal-bearing series
含煤岩系沉积相 sedimentary facies of coal-bearing series
含煤岩系旋回结构 coal-bearing cycle
含煤岩系古地理 paleogeography of coal-bearing series
含煤岩系沉积体系 sedimentary system of coal-bearing series
含煤岩系共生矿产 associated minerals deposits of coal-bearing series
煤层气 coalbed gas, coalbed methane
聚煤作用 coal accumulation process
聚煤期 coal-forming period
含煤区 coal province
煤产地 coal district
暴露煤田 exposed coalfield
半隐伏煤田 semiconcealed coalfield
隐伏煤田 concealed coalfield
含煤性 coal-bearing property
含煤系数 coal-bearing coefficient
含煤密度 coal-bearing density
富煤带 coal-rich zone, enrichment zone of coal
富煤中心 coal-rich center, enrichment center of coal
（聚）煤盆地 coal basin

侵蚀煤盆地 erosional coal basin
塌陷煤盆地 collapsed coal basin, karst coal basin
坳陷煤盆地 depression coal basin
断陷煤盆地 fault coal basin
同沉积构造 syndepositional structure
赋煤构造 coal-preserving structure
煤田预测 coalfield prediction
找煤 search for coal, look for coal
普查 reconnaissance, coal prospecting
详查 preliminary exploration
精查 detailed exploration
找煤标志 indication of coal
煤层露头 outcrop of coal seam
煤层风化带 zone of weathered coal seam
煤层氧化带 zone of oxidized coal seam
勘探方法 method of exploration
勘探手段 exploration means
勘探阶段 exploration stage
勘探区 exploration area
勘探工程 exploration engineering
勘探线 exploratory line
主导勘探线 leading exploratory line
基本勘探线 basic exploratory line
勘探网 exploratory grid
孔距 hole spacing
勘探深度 depth of exploration
勘探程度 degree of exploration
煤田勘探类型 coal exploration type
简单构造 simple structure
中等构造 medium structure
复杂构造 complex structure
极复杂构造 extremely complex structure
煤层稳定性 regularity of coal seam
稳定煤层 regular coal seam
较稳定煤层 comparatively regular coal seam

不稳定煤层 irregular coal seam
极不稳定煤层 extremely irregular coal seam
煤层对比 coal-seams correlation
煤心煤样 coal core sample
筛分浮沉煤样 coal sample for size
瓦斯煤样 coal sample for gas test
地质编录 geological record
煤炭资源量 coal resources
煤炭储量 coal reserves
能利用储量 usable reserves
暂不能利用储量 useless reserves
储量级别 classification of reserves
A级储量 grade A reserves
B级储量 grade B reserves
C级储量 grade C reserves
D级储量 grade D reserves
预测资源量 prognostic resources, predicted resources
远景储量 future reserves, prospective reserves
探明储量 demonstrated reserves, explored reserves
工业储量 industrial reserves
高级储量 proved reserves
保有储量 available reserves
区域地质图 regional geological map
煤田地形地质图 coal topographic geological map
勘探工程分布图 layout sheet of exploratory engineering
钻孔柱状图 borehole columnar section
勘探线地质剖面图 geological profile of exploratory line
煤层对比图 coal-seams correlation section
煤层底板等高线图 coal seam floor contour map
储量计算图 reserves estimation map
煤矿地质 coal mining geology
矿建地质 mine construction geology
生产地质 productive geology
矿井地质条件 geological condition of coal mine
煤矿井地质条件类型 geological condition type of coal mine
煤矿地质勘探 geological exploration in coal mine
煤矿补充勘探 supplementary exploration in coal mine
生产勘探 productive exploration
煤矿工程勘探 coal mine engineering exploration
井筒检查孔 pilot hole of shaft
井巷工程地质 engineering geology in shafting and drifting
瓦斯地质 coalbed gas geology
煤炭自燃 spontaneous combustion of coal
（喀斯特）陷落柱 karst collapse column
探采对比 correlation of exploration and mining information

Unit 2　Rock Mechanics for Mining

矿山岩石力学

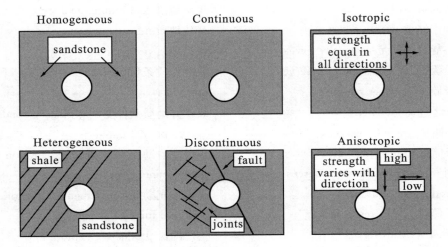

Fig. 2.1　**Diagram of rock structure**

　　Rock mechanics consists of a body of knowledge of the mechanical properties of rock, various techniques for the analysis of rock stress under some imposed perturbation, a set of established principles expressing rock mass response to load, and a logical scheme for applying these notions and techniques to real physical problems.

　　Both the knowledge of the mechanical properties of rock, and the analytical capacity to predict rock mass performance under load, improve as observations are made of in-situ rock behavior, and as analytical techniques evolve and are verified by practical application.

Text 1　Rock Mechanics for Mining　矿山岩石力学

　　Rock mechanics is a field of applied science which has become recognized as a coherent engineering discipline within the last two decades. It consists of a body of knowledge of the mechanical properties of rock, various techniques for the analysis of rock stress under some imposed perturbation, a set of established principles expressing rock mass response to load, and a logical scheme for applying these notions and techniques to real physical problems. Some of the areas where application of rock mechanics concepts have been demonstrated to be of industrial value include surface and subsurface construction, mining and other methods of mineral recovery, geothermal energy recovery and subsurface hazardous waste isolation.

Mining engineering in an obvious candidate for application of rock mechanics principles in the design of excavations is generated by mineral extraction. A primary concern in mining operations, either on surface or underground, is loosely termed "ground control", i. e. control of the displacement of rock surrounding the various excavations generated by, and required to service, mining activity. The particular concern of this text is with the rock mechanics aspects of underground mining engineering, since it is in underground mining that many of the more interesting modes of rock mass behavior are expressed. Underground mines frequently represent ideal sites at which to observe the limiting behavior of the various elements of a rock mass. It should then be clear why the earliest practitioners and researchers in rock mechanics were actively pursuing its mining engineering applications.

In any engineering terms, these are significant accomplishments, and the natural pressure is to build on them. Such advances are undoubtedly possible. Both the knowledge of the mechanical properties of rock, and the analytical capacity to predict rock mass performance under load, improve as observations are made of in-situ rock behavior, and as analytical techniques evolve and are verified by practical application.

In the field of solid mechanics, major advances have been observed in understanding the fundamental modes of deformation, failure and stability of rock under conditions where rock stress is highly in relation to rock strength. In rock engineering practice, the development and demonstration of large-scale ground control techniques has resulted in modification of operating conditions, particularly with respect to maintenance of large stable working spans in open excavations. Each of these advances has major consequences for rock mechanics practice in mining and other underground engineering operations.

The advances in solid mechanics and geo-materials science have been dominated by two developments. First, strain localization in a frictional, dilatant solid is now recognized as a source of excavation and mine instability. Second, variations in displacement-dependent and velocity-dependent frictional resistance to slip are accepted as controlling mechanisms in stability of sliding of discontinuities. Rock-bursts may involve both strain localization and joint slip, suggesting mitigation of this pervasive mining problem can now be based on principles derived from the governing mechanics. The revision has resulted in increased attention to rockburst mechanics and to mine design and operating measures which exploit the state of contemporary knowledge.

The development and deployment of computational methods for design in rock is illustrated by the increased consideration in the text of topics such as numerical methods for support and reinforcement design, and by discussion of several case studies of numerical simulation of rock response to mining. Other applications of numerical methods of stress and displacement analysis for mine layout and design are well established. Nevertheless, simple analytical solutions will continue to be used in preliminary assessment of design problems and to provide a basis for engineering judgment of mine rock performance. Several important solutions for zone of influence of excavations have been revised to provide a wider scope for confident application.

Over those periods, innovations and improvements in engineering practice in mining and mining rock mechanics, and advances in the engineering science of rock mechanics, have been extraordinary. The level of developments in mining rock mechanics science and practice has been as impressive as that in mining engineering. Perhaps the most significant advance has been seen in terms of the resolution of some longstanding issues of rock fracture, failure and strength and their relation to the modes of deformation and degradation of rock around mining excavations.

Some remarkable developments in computational methods have supported these improvements in rock mechanics practice. Many mining rock mechanics problems are effectively four-dimensional, in that it is the evolution of the state of stress over the time scale of the mining life of the ore body which needs to be interpreted in terms of the probable modes of response of the host rock mass. The computational efficiency of tools for three-dimensional stress analysis now permits modeling of key stages of an extraction sequence, for example, as a matter of routine rock mechanics practice. Similarly, computer power and efficient algorithms provide a notable capacity to simulate the displacement and flow of rock in cave mining and to support design of optimum caving layouts.

Not with standing these developments, it is encouraging to note continued attention to formal mathematical analysis in solution of rock mechanics problems. The results of such analysis provide the canonical solutions for the discipline of rock mechanics and ensure a sound base for both the science and engineering practises.

Text 2　Uniaxial Compression Test of Isotropic Rock 各向同性岩石单向抗压测试

Suggested techniques for determining the uniaxial compressive strength and deformability of rock material are given by the International Society for Rock Mechanics Commission on Standardization of Laboratory and Field Tests (ISRM Commission, 1979). The essential features of the recommended procedure are:

- The test specimens should be right circular cylinders having a height to diameter ratio of 2.5~3.0 and a diameter preferably of not less than NX core size, approximately 54 mm. The specimen diameter should be at least 10 times the size of the largest grain in the rock.
- The ends of the specimen should be flat to within 0.02 mm and should not depart from perpendicularity to the axis of the specimen by more than 0.001 rad or 0.05 mm in 50 mm.
- The use of capping materials or end surface treatments other than machining is not permitted.
- Specimens should be stored, for no longer than 30 days, in such a way as to preserve the natural water content, as far as possible, and tested in that condition.
- Load should be applied to the specimen at a constant stress rate of 0.5~1.0 $MPa \cdot s^{-1}$.
- Axial load and axial and radial or circumferential strains or deformations should be recorded throughout each test.
- There should be at least five replications of each test.

Fig. 2.2 shows an example of the results obtained in such a test. The axial force recorded throughout the test has been divided by the initial cross-sectional area of the specimen to give the average axial stress, which is shown graphically against overall axial strain, and against radial strain. The cross-sectional area may change considerably as the specimen progressively breaks up. In this case, it is preferable to present the experimental data as force-displacement curves.

In terms of progressive fracture development and the accumulation of deformation, the stress-strain or load-deformation responses of rock material in uniaxial compression generally exhibit the four stages illustrated in Fig. 2.2. An initial bedding down and crack closure stage is followed by a stage of elastic deformation until an axial stress of σ_{ci} is reached at which stable crack propagation is initiated. This continues until the axial stress reaches σ_{cd} when unstable crack growth and irrecoverable deformations begin. This region continues until the peak or uniaxial compressive strength, σ_c, is reached.

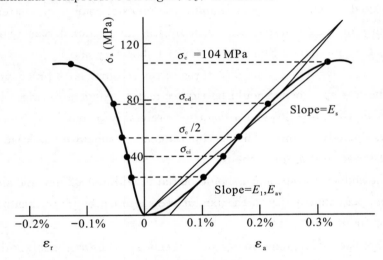

Fig. 2.2 Results obtained in a uniaxial compression test on rock

As shown in Fig. 2.2, the axial Young's modulus of the specimen varies through-out the loading history and so is not a uniquely determined constant for the material.

It may be calculated in a number of ways, and the most common ones are as follows:

• Tangent Young's modulus, E_t, is the slope of the axial stress-axial strain curve at some fixed percentage, generally 50%, of the peak strength. For the example shown in Fig. 2.2, $E_t = 51.0$ GPa.

• Average Young's modulus, E_{av}, is the average slope of the more-or-less straight line portion of the axial stress-strain curve. For the example shown in

Fig. 2.2, $E_{av} = 51.0$ GPa.

- Secant Young's modulus, E_s, is the slope of a straight line joining the origin of the axial stress-strain curve to a point on the curve at some fixed percentage of the peak strength. In Fig. 2.2, the secant modulus at peak strength is $E_s = 32.1$ GPa.

Corresponding to any value of Young's modulus, a value of Poisson's ratio may be calculated as

$$\nu = -\frac{(\Delta\sigma_a/\Delta\varepsilon_a)}{(\Delta\sigma_a/\Delta\varepsilon_r)} \qquad (2.1)$$

For the data given in Fig. 2.2, the values of corresponding to the values of E_t, E_{av}, and E_s calculated above are approximately 0.29, 0.31 and 0.40 respectively.

Because of the axial symmetry of the specimen, the volumetric strain ε_v, at any stage of the test can be calculated as

$$\varepsilon_v = \varepsilon_a + 2\varepsilon_r \qquad (2.2)$$

For example, at a stress level of $a = 80$ MPa in Fig. 2.2, $\varepsilon_a = 0.220\%$, $\varepsilon_r = -0.055\%$ and $\varepsilon_v = 0.110\%$.

In fact, varying the standard conditions will influence the observed response of the specimen. More extensive discussions of these effects are given by Hawkes and Mellor (1970), Vutukuri, et al. (1974) and Paterson (1978).

Terms 常见表达

岩石 rock
岩体 rock mass
容重 unit weight
真密度 true density
孔隙率 porosity
天然含水率 natural moisture content
岩石吸水性 water absorb ability
岩石吸水率 water absorption of rock
岩石膨胀性 rock expansibility
岩石崩解性 rock disintegration
岩石软化性 softening of rock
软化系数 softening coefficient
岩石力学 rock mechanics

矿山压力 rock pressure
矿山压力显现 behavior of rock pressure
岩层控制 strata control
岩体强度 rock mass strength
抗压强度 compressive strength
抗拉强度 tensile strength
断裂强度 fracture strength
抗剪强度 shear strength
杨氏模量 Young's modulus
剪切模量 shear modulus
泊松比 Poisson's ratio
围压 confining pressure
蠕变 creep

弹性变形 elastic deformation
塑性变形 plastic deformation
岩石(体)流变性 theology of rock
岩石(体)非连续性 discontinuity of rock
岩石(体)各向异性 anisotropy of rock (mass)
岩石(体)脆性 brittleness of rock (mass)
岩石(体)膨胀性 expansion of rock (mass)
岩石(体)渗透性 permeability of rock (mass)
岩石(体)碎胀性 rock (mass) bulking, rock swelling
原岩(体) virgin rock (mass), initial rock (mass)
原岩(体)应力 virgin stress of rock (mass), initial stress of rock (mass)
原岩(体)应力场 virgin stress field of rock (mass) initial stress field of rock (mass)
构造应力 tectonic stress
再生应力 redistributed stress, mining-induced stress
围岩 surrounding rock
围岩应力 stress in surrounding rock
应力增高区 stress-concentrated zone
应力降低区 stress-released zone
原岩应力区 initial stress zone, virgin stress zone
应力集中系数 coefficient of stress concentration superimposed
叠加应力 stress superposed
构造裂隙 tectonic fissure
原生裂隙 virgin fissure
采动裂隙 mining induced fissure
岩体水力压裂 hydraulic fracturing of rock mass
岩体水力软化 water softening of rock mass
岩石软化系数 softening factor of rock
顶板来压强度系数 coefficient of roof weighting
冲击载荷 impact load
支承压力 abutment pressure
前支承压力 front abutment pressure
后支承压力 rear abutment pressure
侧支承压力 side abutment pressure
松动压力 broken-rock pressure
变形压力 rock deformation pressure
岩石流变 rheological property of rock
剪切变形 shear deformation
剪切膨胀特性 shear expansion property
单轴抗压强度试验 uniaxial compressive strength (UCS) test
直接拉伸试验 direct tensile test
劈裂试验 splitting test
点载荷试验 point-load test
三轴强度试验 triaxial strength test
直剪试验 direct shear test
三点梁弯曲试验 three-point bending beamtest
现场试验法 field test method
承压板试验 bearing plate method of rock mass deformation measurement
钻孔变形法 drilling deformation method for rock mass deformation measurement
破坏准则 failure criterion
霍克-布朗准则 Hoek-Brown criterion
莫尔-库伦准则 Mohr-Coulomb criterion
莫尔强度理论 Mohr strength theory
格里菲斯强度理论 Griffith's strength theory

原岩应力 in-situ stress
自重应力 gravity stress
残余应力 residual stress
主应力 primary stress
应力释放 stress relief

声发射 acoustic emission
凯泽效应 Kaiser effect
应力解除法 stress relief method
应力恢复法 stress restoration method
水压致裂法 hydraulic fracturing method

Unit 3　Mine Development

矿 井 开 拓

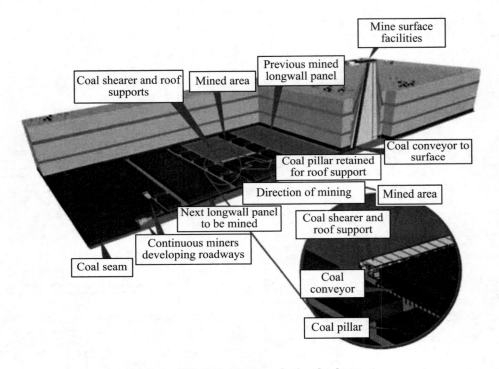

Fig. 3.1　Schematic diagram of mine development

Mine development involves rock excavation of specific objects, to establish the infrastructure. Each object has a designated purpose and becomes a part of the mine's labyrinth of openings. Development objects are vertical shafts, horizontal drifts, inclined ramps and steep raises. Drill-blast rock excavation is the normal way of accomplishment with exception of raises, where the Raise Boring Machine competes with conventional methods.

Unit 3　Mine Development　矿井开拓

Text 1　Mine Field Development　矿区开发

Mine field development involves rock excavation of specific objects, to establish the infrastructure. Each object has a designated purpose and becomes a part of the mine's labyrinth of openings. Development objects are vertical shafts, horizontal drifts, inclined ramps and steep raises. Drill-blast rock excavation is the normal way of accomplishment with exception of raises, where the Raise Boring Machine competes with conventional methods.

1. Shaft Sinking

The shaft is the first component in the development program for the deep mine. The shaft shall, from the start, be excavated to a depth of at least 500 m. The deep shaft is important for the mine's life, to secures many years of production, before ore reserves above the skip station are exhausted.

Extending the shaft in an operating mine is a costly and difficult project which should only occur a mine's very long-range plans. Shaft sinking requires both expert labor and specialized equipment, and is therefore better assigned to a contractor, than done by local personnel. The shaft can be rectangular, circular or elliptical in profile. The circular shaft is simple to excavate and resists rock stresses better than other sections, thus often preferred for a mine's shafts.

2. Drifts and Ramps

Today it is becoming increasingly common to develop a declined ramp for access to the shallow parts of the underground mine. This ramp is used for ore haulage only down to a limited depth since the costs are increasing rapidly. Then ore haulage is made by shaft with skip or inclined conveyor belt. In this way the mine can produce and the cash flow become positive much earlier, thus reducing the risk exposure.

Drifts and ramps build the networks of openings which connects the shaft and workplaces in the mine. Drifts are dimensioned to accommodate machines passing through or operating inside. Space must include a reasonable margin for clearance, walkways, ventilation ducts and other facilities. Cross-sections varies from 2.2 m×2.5 m to 5.5 m×6.0 m, areas from 5.0 m^2 to 25.0 m^2. The 5.0 m^2 drift is enough to accommodate the rail-bound rocker shovel, while the loaded heavy mine truck together with a ventilation duct requires a 25.0 m² transport tunnel.

The grade determines the length of a ramp connecting horizontal levels. Steep grades means short tramming length but are more demanding on equipment. Steep grades should be avoided, there is no reason to save development and sacrifice machines. Normal ramp grades vary from 1∶10 and 1∶7, the steepest grade is 1∶5. The common curve radius is 15.0 m.

A typical ramp runs in loops, with grade 1∶7 on straight sections, reduced to 1∶10 in curves to 1∶10. If the ramp is to be used for trucking up ore from lower levels, the truck manufacturer should be consulted on optimum grade with regards to both capacity and maintenance cost.

Excavating drifts and ramps are standing objects in a mine's development schedule. Development work is often assigned to special teams. The team commands their own equipment plant, with necessary machines, to travel over the whole mine. Advantages of integrated mechanization are exploited.

3. Raises

Raises are steeply inclined openings, connecting the mine's stations at different vertical elevations. The raise may be a ladder-way, for miners to reach their slopes, or an ore-pass to convey ore produced in slopes to haulage levels or an airway in the mine's ventilation circuit. Inclination of raises varies from 55°, the lowest angle for gravity transport of blasted rock, to vertical level. Cross-section of raises vary from 4.0 m² to 6.0 m². The square raise, 2.0 m×2.0 m section, is common.

4. Manual Raise Excavation

Manual excavation of raises is a tough and dangerous job. Yet it has to be done. An example of manual raising is the two-compartment raise, where the miner builds a timber wall dividing the raise into one open and one rock-filled section. The open section is used by the miner to climb the raise by a ladder attached to the wall. Then, standing on top of the rock-filled section, he drills and charges the round above his head. Manual raising is 100% of physical work.

The miner climbs ladders, pulling his rock drill and material to the top of the raise, tied to a rope. He drills the round, charges blast holes lowers his equipment, climbs down and finally triggers the blast. Heights of manually excavated raises are normally limited to 50 m, due to the heavy strains miners are exposed to.

The raise climber eliminates the toughest parts of raising, ladder climbing and pulling materials up by rope. Safety is improved as the platform protects the miner while traveling the raise and drilling from underneath a small canopy. Motorized climbing allows high raises to be excavated, the 100 m raise is no problem for raise climbers. With diesel powered climbing height range exceeds 300 m.

5. Long Hole Raise Excavation

The "drop" raise technique has been applied to short raises, such as breakout raises for sub level open stopping and VCR mining or ventilation raises between sub levels. The raise is long hole drilled all the way, from top to bottom. Blasting made in steps, from bottom upwards. Precision drilling is essential for successful blast result. The long holes can be drilled with the same rigs as for production drilling, which for the mine means that short raises can be excavated without investing in special raising equipment.

It is also possible to drill and blast a slot raise from below with no opening at the top, a "blind" raise. This requires very good practices in both drilling and charging. Still the practical length is limited to 10 m~15 m.

6. Raise Boring

The Raise Boring Machine, RBM is a powerful machine breaking rock with mechanical force. The RBM installs on top of the planned raise and drills a pilot hole, breaking through at the target point at the level below. The pilot bit is then replaced by a reaming bit, with the diameter of the planned raise. The RBM pulls the reaming bit upward, with strong force, while rotating, to break a circular hole in rock.

Text 2 Portal Design 井筒设计

Few items in mine development are as costly as the mine portals since several will be generally employed in a fairly large mine to provide access for men and supplies and for ventilation. Also, specialized equipment and personnel are required for sinking shafts and slopes, and the demand for these is now greater than the supply. However, because portals are the first part of a mine to be installed and usually the last to be abandoned, their usage is long and they require careful design.

1. Site Selection

The selection of a portal site involves numerous factors such as methods of access, methods of ventilation, type of haulage system, and expected life of the mine. That means that certain guidelines need to be followed. The portal area should be free of danger of flooding (low spots) and as close as possible to the geographical center of the property, thus minimizing the cost of such services as power, ventilation, drainage, and haulage. The portal should be located at the lowest elevation of the seam so that the grades favor rail haulage, and natural drainage is away from the working faces to the low point where pumps can be installed to pump the water from the mine to the surface. However, the portal area should be accessible by employees (near highways) as well as close to transportation (rail or barges) with topography suitable for such essential services as facilities as bathhouses, supply yards, preparation plants, refuse disposal, and mine water treatment ponds.

2. Portal Type

Portals are of three types: drift, shaft, and slope. Fig. 3.2(a) is usually an obvious choice, but is becoming increasingly rare as the shallower seams have been depleted. Since it is merely an extension of the underground entry system, it is generally the most economic. However, in certain situations, usually with an irregular property, a drift portal may be so remote that it increases other costs; then it will not be the choice, and a complete cost evaluation becomes necessary.

The decision as to whether a slope or shaft will be employed as a portal can be a very difficult one and reflects the use to which the portal will be put as well as its overall economics. The vertical shaft provides the least length of portal from the surface to the seam. The footage to be sunk and the materials for lining and equipping the shaft are both minimized. In addition, the shaft will provide the least pressure drop which is important for good ventilation.

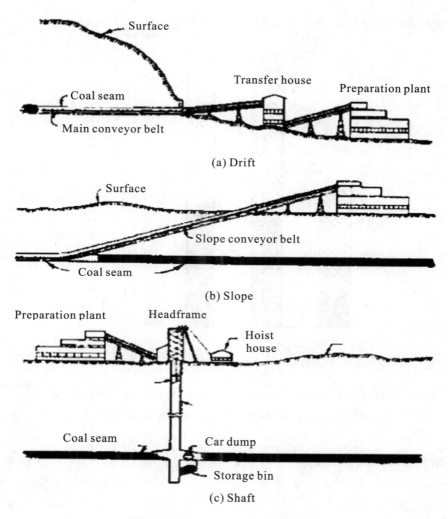

Fig. 3.2 Three types of mine portals

However, haulage, mantrip and supply needs require a careful comparison between shaft and slope portals. A slope configuration is employed to take advantage of belt haulage. Since the maximum angle for belt transportation of coal is 18°, this means that slope length must be roughly three times the depth of the shaft. Fortunately a slope usually can be driven faster than and at approximately one-third the cost per meter of the vertical shaft, offsetting the cost of the greater length required. A deciding factor between shaft and slope can be the condition of the ground. Obviously, if the ground is poor, requiring continuous support, the greater length of the slope will place it at a disadvantage relative to the shaft. Therefore, the slope capital cost may or may not be higher than the shaft depending primarily on ground support considerations.

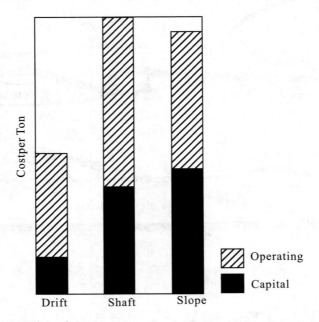

Fig. 3.3　Evaluation of the cost of various types of portals

While vertical hoisting equipment is expensive, slope belts are generally still more costly. This is because of the belt tension requirements of the high vertical lifts. In the USA, the deepest coal belt slope portal is 365 m vertically. Generally, the economics for lifts greater than this are unfavorable. Of course, between 152 m and 298 m, economic considerations will be greatly influenced by the tonnages produced and the life of the mine with large-tonnage long life installations appearing to be economic. The continuous-haulage and large-volume features of a slope belt result generally in very low operating costs. The net effect of these factors can produce the result shown in Fig. 3.3, which compares the three portals for a mine producing very large tonnages with a 30-year life where the ground conditions do not require continuous slope support and the depth is between 152 m and 304 m.

Typical slopes usually have two compartments, Fig. 3.4, either side by side (b) or in the more common over-and-under arrangement (a). In either type, dual provisions for coal belt haulage and rope track haulage are provided. With a slope configuration such as (a), even the largest equipment may be transported into the mine without previously dismantling it, a serious constraint with vertical shafts, especially with the huge machines employed today.

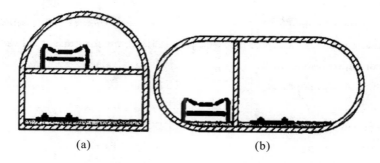

Fig. 3.4 Types of slope configurations

Terms　常见表达

矿区规模 mining area capacity
矿区开发可行性研究 feasibility study of mining area exploitation
矿区总体设计 general design of mining area
矿区地面总体布置 general surface layout of mining area
井田境界 (underground) boundary
井田尺寸 mine field size
矿井井型 production scale of (underground) mine
露天矿规模 surface mine capacity
薄煤层 thin seam
中厚煤层 medium-thick seam
厚煤层 thick seam
近水平煤层 flat seam
缓倾斜煤层 gently inclined seam
中斜煤层 inclined seam, pitching seam
急斜煤层 steeply pitching seam, steep seam
近距离煤层 contiguous seams
矿井可行性研究 (underground) mine feasibility study

露天矿可行性研究 surface mine feasibility study
矿井设计 mine design
露天矿设计 surface mine design
矿井初步设计 preliminary (underground) mine design
矿井施工设计 (underground) mine construction design
矿井设计储量 designed (underground) mine reserves
矿井可采储量 workable (underground) mine reserves
储量备用系数 reserve factor of mine reserves
矿井设计生产能力 designed (underground) mine capacity
煤层产出能力 coal seam productive capacity
矿井服务年限 (underground) mine life
矿井开拓设计 (underground) mine developed design
采区设计 mine-district design, panel design
井田开拓 mine field development

立井开拓 vertical shaft development
斜井开拓 inclined shaft development
平硐开拓 adit development,
　　drift development
综合开拓 combined development
分区域开拓 areas development
阶段 horizon
阶段垂高 horizon interval
阶段斜长 inclined length of horizon
开采水平 mining level, gallery level
辅助水平 subsidiary level
开采水平垂高 lift, level interval
矿井延深 shaft deepening
采区准备 preparation in district
采区 district
分段 sublevel
区段 district sublevel
分带 strip
前进式开采 advancing mining
后退式开采 retreating mining
往复式开采 reciprocating mining
上行式开采 ascending mining,
　　upward mining
下行式开采 descending mining,
　　downward mining
开拓巷道 developing roadway
准备巷道 preparation roadway
回采巷道 entry, gateway, gate
暗井 blind shaft, staple shaft
溜井 draw shaft
溜眼 chute
石门 cross-cut
采区石门 district cross-cut
主石门 main cross-cut
大巷 main roadway
运输大巷 main haulage roadway,
　　main haulage
单煤层大巷 main roadway for single seam
集中大巷 gathering main roadway
总回风巷 main return-roadway
上山 rise, raise
下山 dip
主要上山 main rise
主要下山 main dip
采区上山 district rise
采区下山 district dip
分段平巷 sublevel entry,
　　longitudinal subdrift
区段平巷 district sublevel entry, district
　　longitudinal subdrift
分层巷道 slice drift, sliced gateway
超前巷道 advance heading
区段集中平巷 district sublevel
　　gathering entry
分带斜巷 strip inclined drift
分带集中斜巷 strip mina inclined drift
采区车场 district station, district inset
煤门 in-seam cross-cut
联络巷 crossheading
掘进率 drivage ratio
煤柱 coal pillar
工业广场煤柱 mine plant coal pillar
井田边界煤柱 mine field boundary
　　coal pillar
断层煤柱 fault coal pillar
护巷煤柱 entry protection coal pillar
矿井建设 mine construction
井巷工程 shaft sinking and drifting
矿山地面建筑工程 mine surface
　　construction engineering
矿场建筑工程 mine plant construction
　　engineering

Unit 3 Mine Development 矿井开拓

安装工程 installation engineering
斜井 inclined shaft
平硐 adit
主平硐 main adit
隧道 tunnel
井筒 shaft, slant
井口 shaft mouth
井颈 shaft collar
井身 shaft body
井窝 shaft sump
主井 main shaft
副井 auxiliary shaft, subsidiary shaft
箕斗井 skip shaft
罐笼井 cage shaft
混合井 skip-cage combination shaft
风井 ventilating shaft, air shaft
矸石井 waste shaft
马头门 ingate
井底车场 pit bottom, shaft bottom
环行式井底车场 loop-type pit bottom
折返式井底车场 zigzag-type pit bottom
硐室 (underground) room,
    (underground) chamber
井底煤仓 shaft coal (-loading) pocket
翻车机硐室 tippler pocket, rotary
    dump room
箕斗装载硐室 skip loading pocket
主排水泵硐室 main pumping room
水仓 (drain) sump
吸水井 draw-well, suction well
配水巷 water distribution drift
井下充电室 underground battery-
    charging station
井下主变电硐室 underground central
    substation room
井下(机车)修理间 underground
    locomotive repair room
井下调度 underground control room,
    pit-bottom dispatching room
井下等候室 pit-bottom waiting room
爆炸材料库 magazine
躲避硐 manhole, refuge pocket
腰泵房 stage pump room
巷道 drift, roadway
水平巷道 (horizontal) drift, entry
倾斜巷道 inclined drift
岩石巷道 rock drift
煤巷 coal drift
煤—岩巷 coal-rock drift
人行道 pedestrian way, sidewalk
交岔点 intersection, junction
井筒安全道 escape way
管子道 pipeway
暖风道 preheated-air inlet
检修道 maintaining roadway
单项工程 individual project
单位工程 unit project
矿井建设周期 mine construction period
矿井施工准备期 preparation stage of
    mine construction
井巷过渡期 transition stage shaft to drift
建井期 mine construction stage
矿井建设总工期 overall stage of mine
    construction
矿井建设关键线路 critical path of mine
    construction

Unit 4   Shaft Sinking, Drifting and Roadway Support

井巷施工与支护

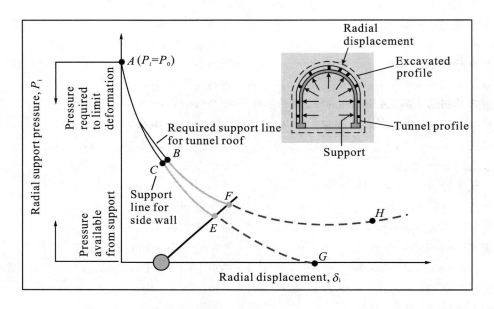

Fig. 4.1   Relationship between roadway deformation and support pressure

Text 1   Shaft Sinking Technology   井筒开凿技术

1. The Progress of Shaft Sinking Technology

Shaft sinking excavation is started from the surface of an opening in the earth. Shafts, which are generally vertical, are usually distinguished from tunnels, which are horizontal. Little difficulty is experienced in shaft sinking through solid rock, which contains little water. When loose, water-bearing strata have to be contended with, careful shoring and sealing of the shaft lining become necessary, and pumping facilities are needed. Shafts are usually circular or rectangular and are generally lined with wood, masonry, concrete, steel, or cast iron. Shafts sunk in loose water-bearing soils, where there is great external pressure on the shaft sides, are nearly always circular; rectangular shafts with wood lining are often used in mining work, as the shafts are frequently of a temporary nature. Shaft sinking through rock is generally accomplished by blasting. When the loose surface material has been removed, holes are drilled, and the charges are placed and are fired by electricity. The broken rock is removed and the process is repeated. In an ordinary rectangular shaft the lining consists of timbers 8 in. or 12 in. (20 cm or 30 cm) square placed horizontally around the shaft. Shafts of a more permanent nature are generally circular in form and lined with cast iron or with concrete masonry 1 ft to 2 ft (30 cm to 61 cm) thick, built in sections as the work advances. When excessive quantities of water are met with, cast-iron tubing is sometimes used. This consists of heavy cast-iron rings made in segments, with flanges for connecting, and bolted together in place. Cement grout is forced into the space between the outside of the tubing and the surrounding earth to form a seal.

Some 120 years ago the invention of the freezing technique was a revolution for the shaft-sinking industry, as it meant that many mineral deposits could now be accessed for the first time through the protective shield of the artificial ice-wall. Freeze sinking still remains the safest way of constructing a mine shaft through unstable, water-bearing ground. The principle itself has hardly changed since those early days, though projects of this type now involve considerable investment in construction and measurement technology, with the deployment of the most up-to-date machines and equipment. The introduction of mobile freeze plant, the transition to rotary drilling with directional turbines for freeze-hole boring and the use of ultrasound measurement technology for checking the freeze-wall thickness are all examples of how the technique has advanced over the years. The freeze-shaft industry has always striven to improve the technology available for calculating the freeze-wall thickness. Drilling and freezing are opposite sides of the same coin, for without the ongoing development of drilling technology the European mining industry would never have been able to sink freeze shafts to the depths that have been achieved. New types of drill pipe, pipe couplings and inner freeze tubes have all made an important contribution in this respect.

2. Shaft Sinking Techniques

(1) Drilling and Firing

The decision as to whether the shaft can be sunk conventionally depends not only on the stability of the ground but also on the assessment of potential water inflow. In the English-speaking countries a water inflow rate of up to one gallon per minute per hundred feet is usually regarded as controllable, which means that in such cases the sinking operation can be carried out with the support of strata injection. Continent countries, by comparison, employ special measures including sealing strips, preliminary injection and stage-freezing to control water up to this magnitude. During the last half-century the shaft-sinking industry gradually moved away from brick work linings and, as in other industries, concrete took over as the dominant construction material.

(2) The Freeze-shaft Method

Shaft construction usually begins with the foreshaft, which is created using normal civil engineering techniques (sheet piling, bored piles, slot walling or caisson). The foreshaft acts as a start-pipe for the subsequent shaft sinking equipment. If the strata are sufficiently stable the shaft can be sunk "manually" by drilling and firing. However, if the ground stay loose or water-bearing it is often necessary to employ freezing in the affected sections or to bore the shaft by virtue of a stabilizing fluid.

(3) Shaft Boring

Deep drilling and shaft sinking are also related to each other. While percussive, ramming or rotary rod drilling with flushing was a rule some 100 years ago, these techniques are no longer used for the most recent surface shafts. Because the overburden can now be controlled very successfully by ground freezing, shafts bored from the surface down tend to be the exception in Europe, although the 8 ft shaft boring at Betws colliery in South Wales again demonstrated to impressive effect on the hidden potential of this system.

When the strata is stable and dry the modern approach is to use conventional large-hole drilling below ground or raise-boring, a technique developed from the American ore mining industry. The use of centrifugal casting to create a load-bearing borehole lining is also a novel and very effective process. Unlike the tunnel construction sector, the shaft sinking industry has still not developed a cost-effective system of rodless full-section excavation from the solid.

(4) Shaft Linings: Brickwork to Concrete

Stable overburden high-cost brickwork linings are no longer used to create the final shaft support structure. Brickwork is now only employed in special cases, and especially for the construction of the external support column when sinking sliding shafts. Even then cast blocks are used, rather than clay bricks. Precast concrete, as used in tunnel construction, has still not gained widespread acceptance and this material is only employed in isolated cases, such as in structural shafts (dewatering shafts and pumping stations) and underground bunkers. Today almost all shaft linings are constructed from poured-in-place concrete of strength category B25 to B45 with nominal reinforcement. Notwithstanding the possible need for static strength, shafts now tend to be designed with wall thicknesses of 30 cm to 50 cm simply for reasons associated with the construction process itself. The lining is often installed in sections using sliding or transferable formwork with ring heights of about 4 m and a small amount of joint clearance, to provide space for the flexibility of the shaft pipes. Since 1977 all surface shafts in the United Kingdom, for example, have been lined with concrete and special techniques and compositions were developed for this purpose during the construction of the Selby Mine Complex between 1977 and 1986. The static calculations were undertaken on the basis of a partially-plastic cross-section and even took account of the thermal impact of the poured in-situ concrete on the frozen ground.

Text 2　Drifting Technology　掘进技术

Drifting operations differ from tunnel operations because of the smaller size, the use and the temporary nature of the drift compared to the tunnel. Drifts are used to gain access to the ore body and for haulage ways. Size varies according to use. Basic equipment and methods which can be combined in various ways, depending upon conditions, for efficient drifting, are summarized in the following.

1. Drills

Proper drilling equipment is important for efficient drifting or raising. Basic considerations in drill selection include type of rock, blasting round to be used, depth of ground, power available, and size of the working area. Air-jacklegs or drifter drills are most often used in drifting and raising. Most drills operate on compressed air, but electric, hydraulic and gasoline-powered units are available.

A drill may be either wet or dry, depending on the method of removing cuttings. Wet drilling is done in hard, abrasive rock; dry-auger in soft rock. A typical air-jackleg drill will cost about $1,500, including the leg. Drifter drills may have either chain or screw feed with separate rotation and feed controls, and a separate water control for wet drilling. Cost is approximately $2,400, plus $1,400 for the feed mechanism.

Multiple drill mounted jumbos are in wide and increasing use. Design variations are plentiful in shape, ranging from one or two booms to much larger two-deck or three-deck jumbos carrying 20 or more automatic-feed drifters. A track-mounted jumbo with two booms will cost approximately $42,400.

Drilling machines with separate rotation motors have proven to be more efficient, and to have higher penetration rates, than conventional units. These can operate, with a screw-feed thrust of 3/4 t to 1.5 t, at 6,000 percussive blows per minute, and 180 rpm to 200 rpm.

2. Blasting

Because of the small headings and the single free face in underground blasting, it is more complicated than open-pit. The type of explosives used, rock properties, and method of loading and sequence of firing the holes, affect hole depth, diameter, alignment and spacing. For these reasons, it is often difficult to formulate exact rules to follow in selecting the optimum blast round. In some cases, experience in similar rock and ore formations is the best guidance in round layout.

A blast round consists of cut, relief, breast, and trim holes. The cut portion is the most important. The objective of the cut is to provide a free face to which the remainder of the round may break.

The two general types of cuts are the angled cut and the burn. These can be used in combinations to form various other cuts. Angled cuts are more advantageous than burn in wide readings due to the fewer holes and less explosive required per foot. A disadvantage is the possibility of large pieces of rock being thrown from the "V".

The wedge or V-cut (Fig. 4.2) consists of two holes angled to meet or nearly meet at the bottom. The cut can consist of one or several Vs, either vertical or horizontal. For deeper rounds or hard-breaking rock, double Vs can be used (Fig. 4.3). The smaller is called the baby cut. It is useful in small headings, since it is located off-center, when lack of room makes it difficult to drill in the center.

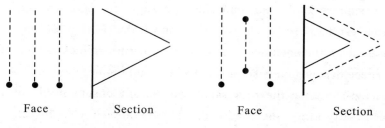

Fig. 4.2 Wedge, or V-cut   Fig. 4.3 V-cut with baby V

Large-diameter burn holes provide excellent relief in big headings. Burn cuts permit deeper rounds than angled cuts and, due to the increased advance per round, may prove more economical. In burn cuts, the holes must be drilled parallelly, with proper spacing, and 0.5 ft to 1 ft deeper than the remainder of the round. Usually, one or more holes (large-diameter) are left unloaded to provide relief for the loaded holes. Various combinations of spacing, alignment and holes loaded are possible, some are shown in Fig. 4.4.

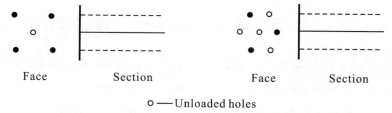

○——Unloaded holes

Fig. 4.4 Burn cuts

3. Mucking

The mucking operation cannot be considered independently from the entire drifting cycle. The type of loading machine must be coordinated with the tramming equipment used. Basic considerations include size of drift, tramming equipment and power available.

The most common mucking machine is the air-operated overcast loader, either track or trackless. There are models with a 45-cu yd hopper that serve as a trammer. Others may have a chain conveyor for loading cars. A typical air-operated track-mounted overcast loader with a conveyor costs around $ 20,000.

Front-end loaders, backhoes, and continuous gathering-loaders also are used in mucking.

4. Support

Proper support is of major importance in a mining operation, since improper support can result in costly delays and repairs. The primary purpose of ground support is to keep the openings wide enough to permit safe extraction of the ore. Support choice depends on cost, ground stress and mining method.

The length of time the rock safely can be left unsupported—bridge action time—will greatly affect the overall drifting-operation cycle. In some cases no support is needed. In others, close support, or support ahead of the last set, is required.

An efficient and adequate way to provide support ahead of the last set is to use pipe spiling. A series of 2.5-in-dia holes, 6-in. spacing, are drilled from behind the last cap at an upward angle of about 20°. Smaller 1.5-in. holes are drilled 1/3 the way down the sides. Two-in. 10-ft-long pipes are driven into the holes drilled across the back and rest on the cap. One-in. pipe is used on the sides. Holes are drilled and pipes are installed before the round is blasted.

Text 3　Roadway Support　巷道支护

Falls of roof account for over 50% of the fatalities that occur in coal mines in the USA. Thus, roof control is one of the more important phases of underground mining. In reality, the control of roof influence the system of mining and is a major determinant of the width and spacing of working places, operations at the mining face, ventilation control, and surface subsidence. Frequently, the control of the roof is the largest single cost item.

Roof control is a never-ending task, not only at the working face and throughout sections where men are working, but along haulways and airways that must be maintained for the life of the mine. This is because roof, even the best of roof will undergo slow deterioration, so it must be frequently examined with corrective support applied when needed.

Each working section of a mine is required to have a roof support plan to provide the minimum protection under most potentially occurred conditions. Such a plan is considered adequate if it does not allow too many failures during the period necessary to extract the coal. It is well to emphasize that this is a minimum requirement that must be supplemented by additional support if loose or badly sagging top is detected, or where unconformities such as clay veins, slips, kettle bottoms, displacement faults, or locally badly fractured roof are discovered.

For support of long-life haulways, air courses, shops, substations, overcasts, track intersections, pump rooms, and other permanent or semi-permanent places, additional support is often required because of the longer working hours factor causing the roof to gradually weaken, or because of spalling caused by warm, humid intake air.

Common methods of main entry permanent support are:

- Additional bolts, or longer bolts in places.
- Posting, or cribs, at breakthrough intersections, with treated timber.
- Additional posts along conveyor lines, sometimes supplemented with crossbars (headers) or half-headers, also with treated timber to prolong life.
- Steel crossbars resting on posts. Along haulage roads a better method is insertion of steel beams or rails into holes drilled in each rib. A good alternative is rail stubs inserted in the holes to support rails along each rib, which, in turn, support steel crossbars. Adequate blocking between roof and crossbar is necessary. Sometimes the roof between bars must be planked or lagged if the roof spalls or breaks in smaller pieces. In case of high roof falls above the coal seam level, cribs with lagging must be erected on the steel crossbars.
- In mines where it is known that the roof will continuously and seriously deteriorate, it is sometimes better and cheaper in the long run to gunite the roof and ribs. Gunite is a mixture of cement, sand, and water blown by air pressure onto the surfaces to be coated. It really provides a 25.4 mm (1 in.) or thicker layer of concrete that seals the air from the moisture-previous rock. Many miles of haulways are so protected in this manner in the USA. It is also valuable in construction of underground shops. Sometimes it is used to line shafts and slopes, either as plain gunite or reinforced with steel mesh. A main requirement to make gunite successful is to scale all loose particles and remove dust before applying.
- Akin to gunite is the "painting" of roof and sides. Here again, the painting is merely a weather seal to prevent warm, moisture-laden air from causing deterioration to the coated surface, as the paint itself has no strength. Various materials have been used in it, varying from a coal tar sealer to some of

the newly developed fireproof synthetic coatings.

• Concerning shaft and slop bottoms it is common to construct concrete walls with or without concrete arches to support wide or potentially troublesome places. Concrete is also used for piers at wide intersections on track haulage, although concrete block walls or wooden cribs are more common.

• Yieldable steel arches are in use at some mines over important haulage roads or in any condition along entries where the roof is hard to support otherwise. When properly installed, and lagged, blocked, backfilled, or cribbed securely between the arches and the roof as well as its ribs, such support is as nearly foolproof as any that can be installed.

Terms　常见表达

井巷掘进 shaft and drift excavation
井巷施工 sinking and drifting
一次成井 simultaneous shaft-sinking
一次成巷 simultaneous drifting
凿井 shaft sinking
井巷工作面 sinking and drifting face
普通凿井法 conventional shaft sinking method
钻眼爆破法 drilling and blasting method
超前水井 pilot shaft
板桩法 sheet piling method
撞楔法 wedging method
掩护筒法 shielding method
全断面掘进法 full-face excavating method
导硐掘进法 pilot drifting method
台阶工作面掘进法 bench-face driving method
宽工作面掘进法 broad face driving method
独头掘进 behind heading
单工作面掘进 single heading
多工作面掘进 multiple heading

贯通（掘进）holing-through
短路贯通 short cut holing of main and auxiliary shafts
特殊凿井法 special shaft sinking method
冻结凿井法 freeze sinking method
冻结器 freezing apparatus
冻结孔 freezing hole, freeze hole
冻结壁 freezing wall
冻结期 freezing period
冻结壁形成期 formable period of freezing wall
冻结壁维持期 maintainable period of freezing wall
冻结壁交圈 closure of freezing wall
冻结压力 pressure of freezing wall, freezing pressure
注浆孔 grouting hole
注浆凿井法 grouting sinking method
上行式注浆 upward grouting
下行式注浆 downward grouting
混合式注浆 multiple grouting
注浆压力 grouting pressure
钻井（凿井）法 shaft drilling method,

sinking by boring method
钻井机进给力 drilling pressure
泥浆护壁 mud off, shaft wall protection by drilling mud
洗井 flushing
泥浆净化 purification of drilling mud
井壁筒 cylindrical shaft wall
固井 consolidation of shaft wall
沉井 caisson
沉井(凿井)法 drop shaft sinking method
震动沉井法 caisson sinking method under forced vibration
淹水沉井法 caisson sinking method in submerged water
套井 surface casing shaft guide-wall
槽孔 trench hole
矸石山 waste dump, refuse heap
装岩 mucking, muck loading
爬道 climbing tram-rail
临时短道 temporary short rail
调车器 car-transfer, car-changer
调车盘 transfer plate, turn plate
井筒装备 shaft equipment
罐道 guide
吊架 hanger
管子间 pipe compartment
梯子间 ladder compartment, ladder way
反向凿井 raising
吊罐 hanging cage
护顶盘 protection stage
支架 support, timber
金属支架 metal support, steel support
砖石支架 masonry support
混凝土支架 concrete support
钢筋混凝土支架 reinforced-concrete support
混合支架 composite support, composite timber
可缩性支架 yieldable support, compressible support
刚性支架 rigid support
完全支架 full support, full timber
梯形支架 ladder-shaped support, ladder-shaped timber
矩形支架 rectangle support
拱形支架 arch support, arch set
马蹄形支架 U-shaped support, horse-shoe set
圆形支架 circular support, circular set
椭圆形支架 ellipse support
背板 set lagging, shuttering
拱碹 arch
碹岔 junction arch, intersection arch
牛鼻子碹岔 ox-nose-like junction arch
砌碹 arching, masonry-lining
井壁 shaft wall, shaft lining
复合井壁 composite shaft lining
支护 supporting
井巷支护 shaft and drift supporting
锚杆支护 bolt supporting
喷浆支护 gunite, guniting
喷混凝土支护 shotcreting
锚喷支护 bolting and shotcreting
锚网支护 bolting with wire mesh
锚喷网支护 bolting and shotcreting with wire mesh
掘进机械 road-heading machinery, driving machinery
巷道掘进机 tunneling machine
井筒掘进机 down-the-hole shaft boring machine
钻井机 shaft-boring machine

钻巷机 drift-boring machine
反井钻机 raise-shaft boring machine
钻装机 drill loader
全断面掘进机 tunnel boring machine, full facer
部分断面掘进机 partial-size tunneling machine
悬臂式掘进机 boom-type road header, boom road header
截割头 cutting head
悬臂 boom
铲装板 apron
可爬行坡度 passable gradient
最小转弯半径 minimal curve radius
离地间隙 ground clearance of machine
钻孔机械 drilling machine
潜孔钻机 down-hole drill
潜孔冲击器 down-hole hammer
探钻装置 probe drilling system
锚杆钻机 roof bolter
风镐 air pick
煤电钻 electric coal drill
岩石电钻 electric rock drill
凿岩机 hammer drill
气腿 airleg
凿岩台车 drill carriage, drill jumbo
钻头 (bore) bit
钻杆 drill rod
钎杆 stem
钎尾 bit shank
装载机械 loader
装煤 coal loader
装岩机 rock loader
耙斗装载机 scraper loader
耙斗 scraper bucket
扒爪装载 gathering arm loader, collecting-arm loader
扒爪 gathering arm, collecting arm
铲斗装载机 bucket loader
铲斗 bucket
铲力 bucket thrust force
侧卸式装载机 side discharge loader
煤巷 coal roadway
半煤岩巷 half-coal and half-rock roadway
锚杆杆体破断力 breaking force of bolt bar
锚杆拉拔力 pulling force of bolt
锚固力 anchor capacity
设计锚固力 design anchor capacity
树脂锚杆 resin anchor bolt
树脂锚固剂 capsule resin
锚固长度 anchorage length
端头锚固 end anchorage
全长锚固 full-length anchorage
加长锚固 lengthening anchorage
拉拔试验 pulling test for bolt anchored by single resin capsule
搅拌时间 stirring time
等待时间 hold time
预紧力 pretension force
预紧力矩 moment of pretension
锚杆快速安装 rapid mounting of bolt
初始设计 initial design
信息反馈 information feedback
正式设计 final design
巷道顶板离层临界值 critical value of roof delamination
复杂地段 section
异常情况 abnormal phenomena
新奥法 new Austrian tunneling method, NATM
爆炸材料 explosive material
岩石炸药 rock explosive

露天炸药 opencast explosive
煤矿许用炸药 permitted explosive for coal mine
爆炸压（力） explosion pressure
爆轰压（力） C-J detonation pressure, detonation pressure
炸药作功能力 explosive power, explosive strength
掏槽眼 cut hole, breaking hole
周边眼 periphery hole, contour hole
辅助眼 satellite hole, easer hole
顶眼 roof hole
底眼 bottom hole

帮眼 flank hole, end hole
直眼掏槽 burn cut, shatter cut
斜眼掏槽 angled cut, oblique cut
混合掏槽 combinational cut, combination burn cut
分阶掏槽 composite cut
装药 charge
起爆 initiation, initiating
起爆药卷 primer cartridge
集中装药 concentrated charge
炮泥 stemming material
爆破 blasting, shooting

Unit 5 Coal Mining Technology and Method

采煤工艺与采煤方法

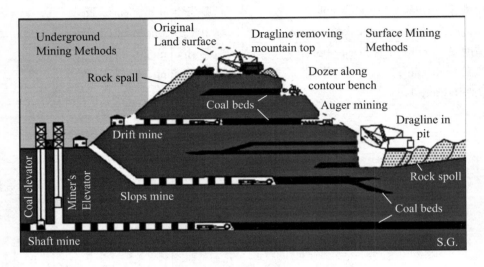

Fig. 5.1 Overview of different mining methods

Text 1 Classification of Mining Operation 采矿作业的分类

Coal mining involves both basic unit operations in the production and handling of coal and auxiliary operations of service functions in nature that are essential but do not contribute directly to the output of coal. These may be classified as follows:

Unit Operations	**Auxiliary Operations**
Cutting	Ventilation
Drilling	Ground Control (roof bolting)
Blasting	Drainage
Loading	Power
Hauling	Communications and Lighting

A mining system is the result of integrated planning for these unit and auxiliary operations and involves determining the most effective deployment of men and machinery. Because of the wide variety of natural conditions encountered underground, different mining systems have been evolved which incorporate the auxiliary and unit operations into effective, overall plans of action. The ultimate objective is, of course, to provide the greatest possible production (in number of tons per unit shift or mine tons per year) and productivity (in number of tons per manshift) consistent with maintaining health and safety.

Frequently, it is inferred in the mining industry that safety and productivity are inconsistent with one another and that the mining system selected must represent a compromise between the two. Actually, the safest system will always prove to be the most productive, and vice versa. Occasionally, however, in changing from a long-practiced undesirable procedure to answer safer one, temporary disruptions and some reduction in production occur. However, over the long term, the safer procedure will soon surpass the unsafe one in efficiency.

It is essential that health and safety procedures be an integral part of the daily operation. Thus, health and safety factors will be considered side by side with technological and economic considerations throughout this book, and, in fact, under the greatest emphasis.

Text 2　Coal Mining Methods　采煤方法

Longwall mining and room-and-pillar mining are the two basic methods of mining coal underground, with room-and-pillar as the traditional method in the USA. Both methods are well suited to extracting the relatively flat coalbeds (or coal seams) typical of the USA. Although widely used in other countries, longwall mining has only recently become important in the USA, its share of total underground coal production having grown from less than 5% before 1980 to about half in 2007. More than 85 longwalls operate in the USA, most of them in the Appalachian region.

In principle, longwall mining is quite simple (Fig. 5.2). A coalbed is blocked out into a panel averaging nearly 800 ft in width, 7,000 ft in length, and 7 ft in height, by excavating passageways around its perimeter. A panel of this size contains more than 1 million short tons of coal, up to 80% of which will be recovered. In the extraction process, numerous pillars of coal are left untouched in certain parts of the mine in order to support the overlying strata. The mined-out area is allowed to collapse, generally causing some surface subsidence.

Extraction by longwall mining is an almost continuous operation involving the use of self-advancing hydraulic roof supports, a sophisticated coal-shearing machine, and an armored conveyor paralleling to the coal face. Working under the movable roof supports, the shearing machine rides on the conveyor as it cuts and spills coal onto the conveyor for transport out the mine. When the shearer has traversed the full length of the coal face, it reverses direction and travels back along the each roof support, the support is moved closer to the newly cut face. The steel canopies of the roof supports protect the workers and equipment located along the face. While the roof is allowed to collapse behind the supports as they are advanced. Extraction continues in this manner until the entire panel of coal is removed.

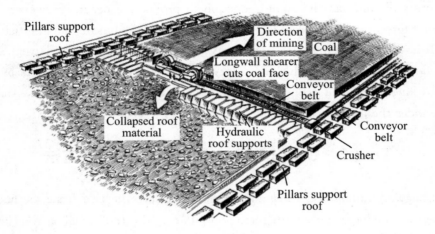

Fig. 5. 2   Longwall mining

Other underground coal mines are laid out in a checkerboard of rooms and pillars(Fig. 5. 3), and the mining operation involves cyclical, step by step mining sequences. The rooms are the empty areas from which coal has been mined, and the pillars are blocks of coal (generally 40 ft to 80 ft on a side) left to support the mine roof. Room-and-pillar mining generally is limited to depths of about 1,000 ft because at greater depths larger pillars are needed, resulting in smaller coal recovery (typically 60% of the coal in the affected area).

In "conventional" room-and-pillar mining, production involves in five steps: mechanically undercutting the coalbed, drilling holes into the bed for explosives, blasting the coal, loading the broken coal into shuttle cars for delivery to a conveyor, and then bolting the mine roof in the excavated area.

To provide a steady flow of coal in a room-and-pillar mine, several stages of mining occur simultaneously in different rooms. A final phase of mining termed "retreat mining" may be performed to recover additional coal by extracting pillars and allowing the roof to fall.

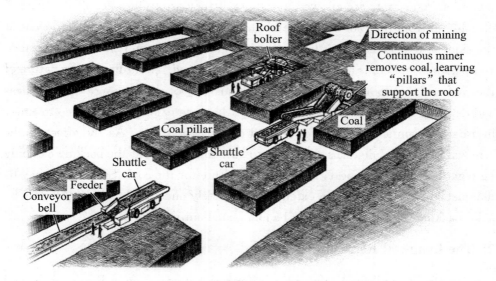

Fig. 5.3 Room and pillar mining

The "continuous" version of room-and-pillar mining is the most common, representing more than half of all underground production. In this method, a continuous mining machine excavates the coal and loads it onto a conveyor or shuttle car in a single step. Despite the term "continuous", the machine operates only part of the working time, because after mining advances about 20 ft, the machine is withdrawn from the face so that roof bolts can be installed to bond the strata and prevent caving.

Text 3　Coal Mining Technology in Longwall Mining
　　　　长壁采煤法

Longwall mining consists of driving one or more gates or entries approximately 91 m to 182 m apart, providing an interconnection and then mining the rib of the interconnection on a longwall, hence the name, longwall mining.

The longwall can be mined either on retreat or on advance.

1. The Longwall Advancing Working

In the longwall system a continuous line of working-face advances in one direction; the face may be straight, curved, or stepped, leaving behind the "goaf" or "gob", in the space previously occupied by the coal seam. In this area of goaf, the either roof strata are allowed to fall, or if they are to be controlled, "packs" of material from the roof strata are built at regular intervals, and in any event at the sides of roads leading to the coalface, along which the mineral is transported, and persons, materials and ventilation may get passed. According to the nature of the roof and the floor, the quality and the quantity of the packing, the thickness and depth of the seam, the roads have usually to be enlarged from time to time, in order to maintain them at the desired cross section area. As the effect of the removal of the coal disturbs the strata to such an extent that the packing can only be regarded, in this respect, as a poor substitute for the coal. Such effect may be diminished by packing to the extent of completely and tightly filling the goaf by solid packing, but they can never be eliminated entirely.

2. The Longwall Retreating Working

As already indicated, the main difference between longwall advancing and longwall retreating is that in the former the coal is transported from the working face along roadways formed in the goaf, whereas in the latter it is transported from the face along roads previously driven through the solid coal which is worked as that face "retreats" towards the shaft bottom. The goaf is left behind and no roads are maintained through it as in the advancing system.

The method is costly in development process as roads have to be driven in the solid to the points where faces are opened out for retreating and much capital expenditure has to be incurred before full production is obtained. In consequence it is not so popular as longwall advancing. Another disadvantage is that it is not so flexible as the mechanized bord and pillar and the longwall advancing methods.

The retreating system, however, calls for little or none of the expenditure usually involved in maintaining roads through the goaf, particularly if careful planning is combined with accurate observation of strata behavior and control. The system has much in its favor, and in many cases would probably show a higher over-all profitability than the advancing system. But it must be recognized that such results would not be achieved in every case, especially where technical information of all relevant facts is not available at the outset.

Much of the success of the system in any circumstances depends upon the distance that the requisite number of roads has to be driven in the solid before the retreating faces can be put into full production, and upon the effect of the natural phenomena upon these roads until they are no longer needed. This involves choosing wisely a boundary, which may be either that of the mine, or a natural or arbitrary one from which to start the longwall retreating work.

The faces may be arranged conveniently in panels and the number of gates required and other factors will, in general, be the same as for longwall advancing, except for the absence of packing in the goaf. Although packing is not carried on as part of the normal retreating system, it may be required where it is necessary to reduce, or attempt to eliminate, the effect of subsidence at the surface caused by the removal of the seam. When the seam is liable to spontaneous combustion it may be necessary, as already mentioned, to seal off the goaf by a regular system of packing, or temporary seals across the open goaf will require to be erected at appropriate intervals.

The longwall system consists of a combination of three basic equipments: a support system; a coal mining machine; a haulage system. There is the arrangement of a long prop-free front face with continuous transportation by a chain conveyor which also acts as a track, or a guide, for a shearing machine. The shearing machine pulls itself back and forth on a stationary chain, cutting slices, or webs, from the face. This protected area is created by cantilevered roof beams that extent from powered supports. By providing a longer and more substantial cantilever frame, this one-web-back system permits the machine operator to travel between the conveyor and the front props.

As has noted at the beginning of this chapter, longwall is not new in the USA, having been used through several periods since before 1900. Invariably, the method appeared uneconomical when compared to the room-and-pillar method and was abandoned in favor of it. Basically, labor costs incurred in moving the manual supports were just too high. Since the introduction of self-advancing hydraulic props in 1960, longwall gained considerable popularity with the promise of much greater application yet to come. Because roof pressures usually make it possible to maintain only a limited cantilevered top, the actual working space is generally limited from 3 m to 3.6 m and the machine operator and other workers must travel in the confined area between props. Designing highly productive equipment to fit such limited space, maintaining it properly, and providing for adequate movement of men are no simple tasks. Obviously, the space for the mining machine, conveyor, and supports must be kept as narrow as possible. But since the web mined with each pass varies from as little as 5.1 cm (2 in.) with some planers to as much as 91 cm (36 in.) with some shearers, the equipment must be capable of being moved quickly as the face retreats.

Because the conveyor must be snaked, the flexibility that only an armored-chain type can provide is called for. However, in addition to the usual problems of chain conveyors, the long lengths of them required in longwall mining result in other difficulties. Faces that have reached 274 m in width have invariably had ad chain problems. Until recently, the chain conveyor was the weakest link in the whole longwall system, limiting the length of the longwalls, restricting the productivity of high-capacity mining equipment. However, heavy-duty conveyor with improved low-profile tail drives and single or dual center chains of increased wire size have been introduced in recent years with the result that the face conveyor is no longer such a weak link in the total system.

The coal wining machine can be located between the conveyor and the face, resting on the bottom, or it can be mounted so that it rides on top of the conveyor. Since the planer is narrow and takes a thin slice, it rests on the bottom. Shearers until recently were always conveyor mounted, decreasing conveyor life, but in-the-web shearers have now been introduced. Chain conveyor used with planers have about five times the life of those used with shearers. With either machine, coal is now mined in both direction or bidirectionally.

There are various configurations of longwall supports but all are hydraulically operated, have a number of legs (props or jacks) with roof and bottom bars, and are self-advancing by using a double-acting hydraulic ram. Whereas prop load capacities were only 9.1 ft to 18 ft each just a few years ago, they are now up to 18 ft. Support components will be described in detail later.

Terms　常见表达

采煤方法 coal mining method, coal mining method
回采工作面 coal face, working face
采煤工艺 coal mining technology

长壁工作面 longwall face
短壁工作面 shortwall face
双工作面 double (-unit) face
对拉工作面 double (-unit) face

Unit 5　Coal Mining Technology and Method　采煤工艺与采煤方法

煤壁 wall
采高 mining height
开切眼 open-off cut
工作面端头 face end
切口 stable, niche
始采线 beginning line, mining starting line
终采线 terminal line
采空区 goaf, gob, waste
工作面运输巷 headentry, headgate, haulage gateway
工作面回风巷 tailentry, tailgate, return airway
破煤 coal breaking, coal cutting
爆破采煤工艺 blast-mining technology
普通机械化采煤工艺 conventionally-mechanized coal mining technology
综合机械化采煤工艺 fully-mechanized coal mining technology
掏煤 slotting
爆破装煤 blasting loading
循环 working cycle
循环进度 advance of working cycle
旋进式推进 revolving mining turning longwall
跨采 over-the roadway extraction
整层开采 full-seam mining
分层开采 slicing
走向长壁采煤法 longwall mining on the strike
倾斜长壁采煤法 longwall mining to the dip or to the rise
倾斜分层采煤法 inclined slicing
短壁采煤法 shortwall mining
煤房 room, chamber
房式采煤法 room mining, chamber mining
房柱式采煤法 room and pillar mining
掩护支架采煤法 shield mining
伪倾斜柔性掩护支架采煤法 flexible shield mining in the false dip
倒台阶采煤法 overhand mining
正台阶采煤法 heading-and-bench mining
水平分层采煤法 horizontal slicing
斜切分层采煤法 oblique slicing
仓储采煤法 shrinkage stoping
伪斜长壁采煤法 oblique longall mining
长壁放顶煤采煤法 longwall mining with sublevel caving
水平分段放顶煤采煤法 top-slicing system of sublevel caving
放煤步距 drawing interval
放煤顺序 drawing sequence
放采比 drawing ratio
自重充填 gravity stowing
机械充填 mechanical stowing
风力充填 pneumatic stowing
水力充填 hydraulic stowing
注砂井 storage-mixed bin, sandfilling chamber
充填步距 stowing interval
充填能力 stowing capacity
充采比 stowing ratio
充填沉缩率 setting ratio
采煤机械 coal wining machinery, coal getting machinery
截煤机 coal cutter
截装机 cutter-loader
刨煤机 plow, plough
采煤机 coal winning machine
滚筒采煤机 shearer, shearer-loader
钻削采煤机 trepanner

连续采煤机 continuous miner
骑槽式采煤机 conveyor-mounted shearer
爬底板采煤机 off-pan shearer, floor-based shearer, in-web shearer
普通机械化采煤机 conventionally-mechanized coal wining face unit
综合机械化采煤机组 fully-mechanized coal winning face unit
采煤联动机 coal-face wining aggregate
机面高度 machine height
过煤高度 underneath clearance, passage height under machine
截深 cutting height, web depth
下切深度 dinting depth, undercut
调高 vertical steering
调斜 roll steering
工作机构 working mechanism, operating organ
截盘 cutting jib, cutting bar
截链 cutting chain
截深滚筒 cutting drum
螺旋滚筒 screw drum, helical vane drum
钻削头 trepan wheel
截齿 pick, bit
截线 cutting line
截距 intercept
截齿配置 lacing pattern
截槽 kerf
切槽 cutting groove
截割速度 cutting speed
切削深度 cutting depth
截割阻抗 cutting resistance
截割比能耗 specific energy of cutting
磨砺性系数 coefficient of abrasiveness

截割部 cutting unit
摇臂 ranging arm
行走部 travel unit, traction unit
行走驱动装置 travel driving unit
行走机构 travel mechanism, traction mechanism
牵引力 haulage pull, tractive force
牵引速度 haulage speed, travel speed
内牵引 internal traction
外牵引 external traction
机械牵引 mechanical haulage
液压牵引 hydraulic haulage
链牵引 chain haulage
无链牵引 chainless haulage
牵引链 haulage chain
内喷雾 internal spraying
外喷雾 external spraying
拖缆装置 cable handler
安全绞车 safety winch
上漂 climbing
下扎 dipping, penetration
进刀 sumping
落道 derailment
静力刨煤机 static plough
动力刨煤机 dynamic plough
拖钩刨煤机 drag-hook plough
滑行刨煤机 sliding plough
滑行拖钩刨煤机 sliding drag-hook plough
刨头 plough head
刨链 plough chain
刨刀 plough cutter
拖板 base plate
调向油缸 lifting ram
前牵引 front haulage
后牵引 rear haulage
刨削深度 ploughing depth

刨削速度 ploughing speed
刨削阻力 ploughing resistance
低速刨煤 low-speed ploughing
高速刨煤 high-speed ploughing

双速刨煤 dual-speed ploughing
定压控制 fixed-pressure control
定距控制 fixed-distance control

Unit 6 Rock Pressure and Strata Control

矿山压力及其控制

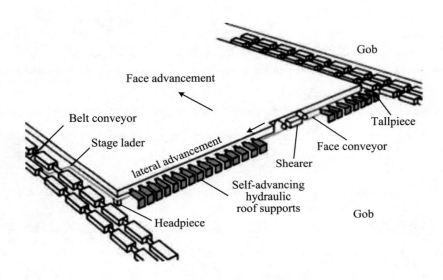

Fig. 6.1 Equipment layout of longwall mining face

Text 1 Rock Pressure in Mine 矿压显现

Conceptually, a softer coal seam is sandwiched between the relatively stronger roof and floor rocks, which are loaded by the weight of the overburden. Stress is uniformly distributed in the coal seam under such conditions. When coal is extracted in slices as it is in the longwall mining, the roof over the gob (waste) area thus created tends to deflect as a cantilever and will eventually cave in. As a result, stress conditions in the longwall panel will be read as balanced until a new equilibrium is achieved. Vertical abutment pressures occur along the edges of the gob area as shown in Fig. 6.2. The abutment pressure occurring in solid coal in front of the longwall face is called the front abutment; pressures at both ends of the face and along the ribs of the headentry and tailentry are the side abutments;

and the rear abutment is located in the gob area bridging between the headentry and tailentry. The side and front abutments intersect at the T-junctions of the head and tailentries and superimpose to become peak abutment pressures.

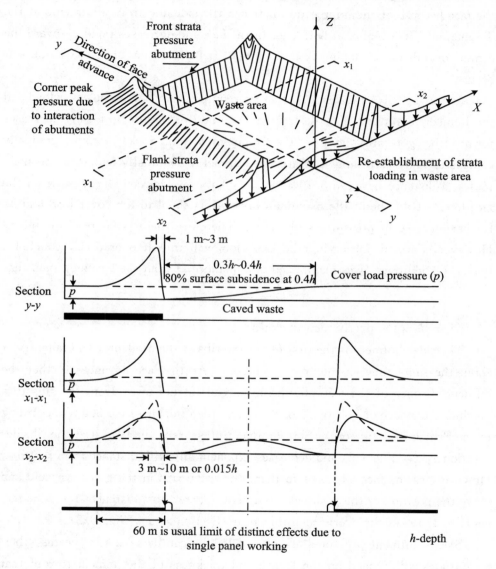

Fig. 6.2 Abutment pressures around a longwall panel

Depending on local conditions, the front abutment is first felt as far as 500 ft outby the face. However, the increase in magnitude at this time is small. It picks up slightly at a distance between 200 ft and 150 ft out by the face, increases rapidly when the face is between 70 ft and 50 ft away, and reaches peak abutment pressure 3 ft to 16 ft ahead of the face (section y-y of Fig. 6.2). The magnitude of the peak abutment pressure ranges from 1.5~5 times the cover load (overburden weight). The magnitude of the front abutment is non-uniform along the face because of the intersection between the side and front abutments at the T-junctions. It is higher at both ends of the face and decreases rapidly toward the center for a distance ranging from 30 ft~70 ft before a uniform front abutment occurs.

At the face area the vertical pressure reduces to far below the cover load and the immediate roof is in a destressed condition. Pressure gradually increases toward the gob area because the fragmented roof rocks become gradually compacted and, according to British researchers, eventually supports the main roof at a distance between 3/10 and 4/10 of the overburden thickness from the face line. At this point the maximum pressure is equal to the cover load and no further increase in pressure is observed. Therefore there is no rear abutment. However, Carmen, observing the USA longwall retreat operations, concluded that a rear abutment was maintained at a distance D from the faceline, such that

$$D = \frac{3h}{20} + 60$$

where h is the overburden thickness.

The side abutment is the first felt at the ribs of the headentry and tailentry at about the same time as the front abutment. As the face advances further the influenced zones of the side abutments extend further away from the ribs to a maximum distance of 1/4 to 1/3 of the overburden thickness (or 200 ft, according to the British investigators). Side abutment pressure is the largest at the ribsides (section x_1-x_1, Fig. 6.2) and decreases exponentially with distance from the ribs. However, as the face advances further, the entry ribs in the gob area yield and move the position of the peak side abutment outward to a distance approximately equal to 1.5% of the overburden thickness (Section x_1-x_1, Fig. 6.2).

Since abutment pressures are caused by the cantilevered roof beams, their magnitudes will depend on the length and thickness of the immediate roof that overhangs the gob area. The longer and thicker the immediate roof, the larger the abutment pressuresare. This is true if a massive sandstone layer is immediately above the coal seam. A weak shale will break immediately behind the powered supports without any overhang.

Strata behavior can be classified into three distinctive regions, the gob, the face, and the solid coal.

1. Gob Area

As the mining proceeds, the gob is left open and no roof caving occurs in the initial [Fig. 6.3(a)]. The width of the empty gob opening continues to widen as the mining progresses. When the empty gob becomes too wide the immediate roof starts to slip, sag, and separate [Fig. 6.3(b)] and eventually it caves in [Fig. 6.3(c)]. Thereafter, the root continues to cave as the supports move forward. Depending on the characteristics of the immediate roof, the distance between the barrier pillar and the face location of the first cave-in in the gob ranges between 56 ft and 400 ft.

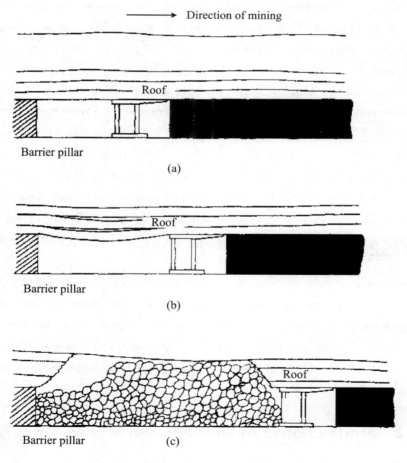

Fig. 6.3 Idealized progress of first caving in a longwall panel

Immediately before the first cave-in, the abutment pressure at the face area reaches the maximum value. This causes the first "weighting" of the face powered supports and may damage the supports if sufficient resistance is not provided. After the first cave-in, the immediate roof generally caves as the supports are advanced, although some types, such as sandstone, tend to hang over the rear edges of the supports for some distance before they collapse. Loud noise and winds are generally associated with cave-ins.

2. Face Area

At the face, the immediate roof supported by the powered supports converges and at the same time moves laterally toward the gobs. Since there is a large frictional force between the roof canopy of the powered supports and the immediate roof, the lateral component of the movement tends to separate the roof into blocks and rotate toward the gobs.

Convergence at the face can be due to floor heaves, coal squeezes, or roof sags. Regardless of its source, convergence at the face is directly related to the support resistances. Considerable measurements done on convergence in European coal fields have indicated that the roof to floor convergence is the same whether it is near the face or at the gob edges. This implies that the immediate roof behaves like a solid slab cantilevered in a soft support at the coal face. Data obtained so far indicate that face convergence per unit time is linearly proportional to the rate of face advance or 0.5 in. convergence per foot of face advance.

3. Solid Coal

When presented with the fact that the immediate roof is weak and overburden pressure is large, on its observations over the solid coal ahead of the longwall faces, it will induce fractures of various types will be induced in the solid coal. Induced cleavages, tiny and largely vertical, have been found some 87 ft while major fractures about 46 ft ahead of the face. Fewer fractures were noted for shorter faces.

Text 2　Strata Control in Underground Mining　矿压控制

With the advent of modern coal mining techniques, it has become imperative to adopt roof bolting as a primary means of support in place of the traditional supports. About 2,500 million of coal has been locked in pillars of which only about 1,000 Mt is amenable to opencast mining, about 1,500 Mt is to be extracted by underground mining. Strata control management is one of the major reasons for losing of pillars.

Although technology has improved nowadays with the introduction of Blasting Gallery method, Integrated Caving method and Hydraulic Mining, some of them are unsuccessful for the failure of trials at Churcha, Kottadih, etc., and many more due to lack of suitable strata control techniques. Salient features that lead to typical problems in underground coal mining include:

• Steeply dipping, faulted, folded, highly gassy beds under aquifers and protected land have remained virgin.

• Developed pillars under fires, surface features sterilized because of acute shortage of sand.

• Development has been in multi sections.

• Highly stressed zones have been created due to barriers causing difficulty of undermining of the seams.

Hazardous roof conditions identified in some mines in other countries were positively correlated with mining activities beneath stream valleys (Mucho and Mark, 1994). Evidence of valley stress relief was found beneath several valleys in the form of bedding plane faults and low-angle thrust faults. At many places the ratio of horizontal to vertical stress was in the range of 2 to 3. This type of failure, previously believed to be only a shallow phenomenon, was also found at increased mining depths.

Horizontal stresses affect a number of USA coal mines (Mucho, et al., 1995). To address the effects of the stress field and to control its potentially damaging effects, a number of control strategies have been developed, such as reorientation of the retreat direction, stress shadowing of the key openings, and altering the mining cut sequence.

However, many of these techniques are highly direction-oriented, and to be effective, they require precise determination of the major (maximum) principal horizontal stress direction. For these types of typical geomining condition associated with high horizontal stress, a system of roof truss was used successfully. Cable bolts were effectively utilized for strata control in thick seams and adverse roof conditions in Indian coal mines (Jayanthu and Gupta, 2001). Roof slotting is one tactic to stress shadow the adjacent workings (Frank, et al.,

1999).

The studies by the National Institute for Occupational Safety and Health (NIOSH) and the Mine Safety and Health Administration (MSHA) at Sargent Hollow Mine, in Wise County, VA, indicated that the weak floor strata was being subjected to, and damaged by, high horizontal stresses. After the "advance and relieve mining method" was implemented, the overall mining conditions at the mine improved, and the roof control plan was approved for further use. Various approaches such as empirical, observational, analytical required for design of strata control system is presented in Fig. 6.4.

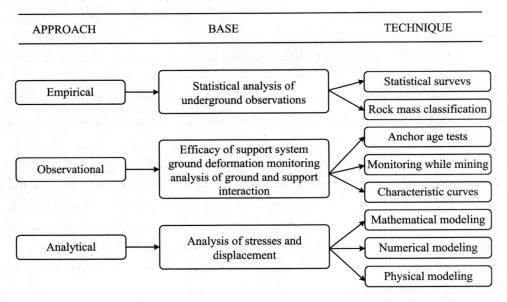

Fig. 6.4 Various approaches for development of strata control techniques

Roof falls have been usually attributed to bad roof strata and high vertical stresses related to the overburden depth. However, for shallow depth conditions, roof falls in recent studies have been attributed to high horizontal stresses (Yajie and John, 1998). About 73 roof falls analyzed in the USA, 37 occurred in the entries with 52° angle with the major horizontal stress. Unfortunately, majority of the locked up pillars in thick seams have strata control problems possibly due to high horizontal stresses, and need careful review of history of falls. Through the horizontal stress recognition features, some of the following control techniques can be effectively implemented:

• Reorienting the drivage direction of the mine openings.

- Panel orientation and retreat direction.
- Stress shadowing through key openings.
- Altering mining cut sequences.

In the light of experience of past few years, the norm for designing of Systematic Support Rules in development roadways needs reexamination and modification. Life of the roadway should also be considered while designing the system. The galleries of a Bord and Pillar system may be self-supporting under a very strong roof or the immediate roof may be supported by props, roof bolts or roof stitching depending on the local conditions (Mathur, 1999). The weight of the main strata is borne by coal pillars. During pillar extraction, props, cogs, roof bolts have been conventionally used in splits, slices, etc., with skin-to-skin chocks near goaf edges (Mobile Roof Supports in the USA). Performance of the support systems have been extensively studied worldwide for understanding the strata mechanics, higher bond strengths and anchorage capacities with reduced annulus was indicated (Tadolini, 1998).

Developments in support systems are related to material for bolts (cuttable bolts, tendon, resin, etc., for grouting, truss bolts), mobility of supports (mobile roof support), capacity of supports (high capacity shields, props) (Gupta and Prajapati, 1997; Khan and Hassani, 1993).

Mobile supports have been successfully deployed for depillaring (Larry, 1998). Support capacities up to 800 t are available and need introduction in Indian coalfields. It provides an upward active force on the immediate roof strata and results in normal cave line pushed back into goaf. This allows a wide stook to be mined while depillaring, thereby costly and relatively unproductive cycle of splitting of pillars and associated support can be minimized. In near future, the concept of man-less mining needs to be adapted to the maximum possible extent for improved safety, production, and productivity.

Continuous monitoring of strata behavior in terms of convergence of openings in advance on either side of the extraction line, and stress levels over pillars, stooks in advance of the extraction and ribs in the goaf was required through remote monitoring instruments for understanding the strata mechanics at critical conditions of roof falls. Continuous monitoring of support pressures was attempted to investigate the rock mass response to mechanized pillar extraction (Follington and Huchinson, 1993).

Text 3　Supports in Longwall Mining　长壁开采支架

If good roof control is the key to successful longwall mining, then the selection of the optimum hydraulic powered support system is of utmost importance to the mine operator. Increased safety, better ground control, increased production, optimum capital outlay, and low operating costs are at stake.

Various methods have been suggested to estimate support requirements for achieving good roof control. These range from computation of the mean load density based on actual measurements to estimates based on the height of caving above a coalbed.

There are several basic types of supports available today: a chock configuration that is quite different from crib-type construction, a frame type which is an adaption of two or more props in line with a beam, two-leg shield supports that are the most recent type introduced in this country, and the four-leg shield or chock-shield which combines the best feature of the chock and shield into a single unit. Within each basic type, there are many variations.

1. Chock Support

The chock is the most commonly employed powered support in use today. Chocks consists of from two to eight hydraulically set legs, box type steel canopy or canopies, single or double base plates, push-pull cylinders, and control valves to lower, advance, and reset the unit, Fig. 6.5(a). The legs are topped with articulated or rigid canopies, and mounted on single or interconnected base plates. In the articulated design, the forward portion of the canopy is hinge-linked to the rear section to provide greater contact with irregular roof. Some of the canopies are designed to permit the use of a pullout cantilever or extension bar to contact the roof ahead of the forward canopy tip. Connection between the legs and canopies is frequently through a ball-and-socket joint for greater flexibility and adaptability. Available in a variety of legs, capacities, and canopy arrangements, chocks offer mighty strength.

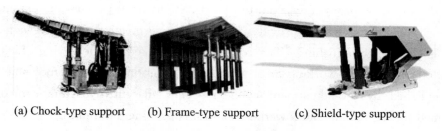

(a) Chock-type support   (b) Frame-type support   (c) Shield-type support

Fig. 6.5 Basic types of supports

2. Frame-Type Support

A frame-type support may consists of two or three jacks grouped in line independently or connected together flexibly in a double frame unit, each frame being capable of independent movement, Fig. 6.5(b). This type of support is usually employed with planers. The two frames are connected with a push-pull hydraulic ram, allowing the right unit to be pushed out using the left frame as an anchor, and then the left frame is pulled up even with the right frame after the right frame is anchored in place. An independent ramming system pushed the conveyor over from the props. Thus a frame support can walk independently of any anchor source. Until the recent advent of shield support, the trend has been from frame to chock supports until today approximately 30% of the longwall supports are the frame type, 65% the chock type, and the remaining 5% the shield type.

3. Shield-Type Support

Shield-type supports (two-leg) were introduced in the USA in the mid-1970s but had been used for many years in Eastern Europe, and since 1970, in Western Europe. However, the response to them in the USA appears so enthusiastic that many experts are predicting shields may eventually capture the entire domestic longwall support market.

A shield-type unit consists of the floor beam, the caving shield, the roof beam, hydraulic valves and hoses, Fig. 6.5(c) and Fig. 6.6. The floor beam provides a large contact area making the shield support suitable even for the soft floors. The advancing ram is connected to conveyor by push-pull rods. Hydraulic legs with yield loads of 100, 131, 158, and 181 t are available. The setting load of the legs is adjustable up to 80% of the yield load. The caving shield, linked to the floor beam by a heavy-duty, pin-type connection, incorporates an extensible angle-shaped plate which provides skin-to-skin protection between adjacent shields against gob flushing. The roof shield is linked to the caving shield to completely cover the working area. Horizontal positioning of the roof shield is accomplished by an adjustment cylinder. The adjacent hydraulic control is designed for leg raising and lowering, and rams pushing and pulling.

Shields have many theoretical advantages of the shields, and some disadvantages, too. The advantage of shields is that they offer a high degree of safety in that they isolate the worker from the gob. There is less dust or rock falling down between the supports and the workers feel more secure. Then, too, because the shields are three-point devices, they are mechanically more rigid than chock and are also less prone to problems from lateral loadings.

Shield can be moved forward under a brushing load. This means that they maintain contact with the roof while they are being moved instead of being lowered and advanced under no load. Because they are constantly in contact with the roof, there are no problems with relaxing rear part of the roof as they are moved forward.

A major disadvantage of shield is their cost. Because they are heavier, they cost more, which means their cost is somewhat proportional to their weight, as with all equipment. Another weakness against them is their lower support capability which is the result of their limited number of legs. The amount of support that can be applied to the roof over the roof plate is generally less than that obtained with chocks.

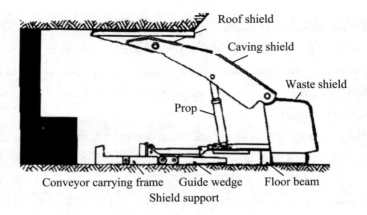

Fig. 6.6 Shield-type units

Terms 常见表达

矿山压力 rock pressure in mine
矿山压力显现 strata behaviors
原岩（体）virgin rock（mass）
围岩 surrounding rock
原岩体应力 initial stress, stress in virgin rock mass
采动应力 mining-induced stress
应力增高区 stress-concentrated area
应力降低区 stress-relaxed area
叠加应力 superimposed stress
自重应力 gravity stress
构造应力 tectonic stress
支承压力 abutment pressure
前支承压力 front abutment pressure
后支承压力 rear abutment pressure
侧支承压力 side abutment pressure
松动压力 broken-rock pressure
变形压力 rock deformation pressure
顶板 roof
底板 floor
伪顶 false roof
直接顶 immediate roof, nether roof
基本顶 main roof
顶板稳定性 roof stability
坚硬岩层 strong stratum, hard stratum
松软岩层 weak stratum, soft stratum
破碎顶板 fractured roof, friable roof
人工顶板 artificial roof
再生顶板 mat, regenerated roof
上覆岩层 overlying strata
离层 bed separation
自然平衡拱 dome of natural equilibrium, natural arch
原生裂隙 initial fissure
构造裂隙 tectonic fissure
采动裂隙 mining induced fissure
岩石软化系数 softening factor of rock
普氏系数 Protodyakonov coefficient
岩石内聚力 rock cohesion
岩石内摩擦角 internal friction angle of rock
岩层控制 strata control
（工作面）顶板控制 roof control
垮落法 caving method
充填法 stowing method
缓慢下沉法 gradual sagging method
煤柱支撑法 pillar supporting
回柱 prop drawing
放顶 caving the roof
初次放顶 initial caving
无特种柱放顶 caving without specific props, caving without breaker props
强制放顶 forced caving
控顶距 face width
放顶步距 caving interval
端面距 tip-to-face distance
无支柱距 prop-free front distance
冒顶 roof fall
顶板破碎度 roof flaking ratio
顶板单位破碎度 specific roof flaking ratio
局部冒顶 partial roof fall
区域性切冒 extensive roof collapse
压垮型冒顶 crush roof fall
推垮型冒顶 thrust roof fall
端面冒顶 roof flaking (in tip-to-face area)
漏顶 face roof collapse with cavity
片帮 rib spalling, sloughing

顶板垮落 roof caving
顶板垮落角 roof caving angle
不规则垮落带 irregularly caving zone
规则垮落带 regularly caving zone
岩石碎胀系数 bulking factor, swell factor
垮采比 caving-height ratio
顶板压力 roof pressure
初次来压 first weighting
周期来压 periodic weighting
动载系数 dynamic load coefficient
顶底板移近量 roof-to-floor convergence
顶底板移近率 roof-to-floor convergence ratio
顶板回弹 roof rebound
顶板台阶下沉 roofstep
顶板弱化 roof weakening
煤/岩固化 coal/rock reinforcement
底鼓 floor heave
冲击地压 rock burst, pressure bump
矿震 shock bump
恒阻支柱 yielding prop
增阻支柱 late bearing prop
摩擦支柱 frictional prop
单体液压支柱 hydraulic prop
迎山角 prop-setting angle
特种支柱 specific prop
放顶(支)柱 breaker prop
墩柱 heavy-duty pier, hydraulic breaker prop
支垛 crib
十字顶梁 cross bar
滑移顶梁支架 slipping bar composite support
柔性掩护支架 flexible shield support
气囊支架 air-bag support

支柱密度 prop density
初撑力 setting load
初撑力强度 setting load density
(额定)工作阻力 yield load, working resistance
有效支撑能力 practical supporting capacity
支护强度 supporting intensity
支护刚度 support rigidity
支撑效率 supporting efficiency
支架可缩量 nominal yield of support
循环平均阻力 time-weighted mean resistance, mean load per unit cycle
压入强度 press-in strength
底板载荷集度 floor load intensity
围岩稳定性 stability of surrounding rock
静压巷道 workings subject to static pressure
动压巷道 workings subject to dynamic pressure
巷道断面缩小率 roadway reduction ratio
无煤柱护巷 non-chain-pillar entry protection
沿空巷道 gob-side entry
沿空留巷 gob-side entry retaining
锚梁网支护 roof bolting with bar and wire mesh
巷旁支护 roadside support
巷旁充填 roadside packing
矸石带 waste pack, strip pack
架后充填 backfill
挑顶 roof ripping
挖底 floor dinting
最低高度 closed height
伸展高度 extended height
支护阻力 support resistance

Unit 6　Rock Pressure and Strata Control　矿山压力及其控制

推移步距 advance increment, mm
架宽 unit installation spacing
底座类型 base type
油缸数量 number of hydraulic legs
立柱类型 leg type
油缸活塞直径 piston diameter, mm
阀压力 check valve actuation pressure
工作阻力 working pressure, MPa
额定加载率 setting load ratio
平均工作阻力 average ground pressure
推移模式 advancing pattern
推移力 advance force (nominal)
支架结构 support unit
传输机 conveyer (a pan)
油缸服务年限 service life of steel works, cycle
支架控制类型 support control type
液压支架 hydraulic support, powered support
支撑式支架 standing support
垛式支架 chock support
节式支架 frame support
架节 support unit, support section
掩护式支架 shield support
支撑掩护式支架 chock-shield support
端头支架 face-end support
锚固支架 anchor support
迈步式支架 walking support
即时前移支架 immediate forward support, one-web back system
放顶煤支架 sublevel caving hydraulic support
铺网支架 support with mesh-laying device
最大结构高度 maximum constructive height
最小结构高度 minimum constructive height
最大工作高度 maximum working height
最小工作高度 minimum working height
支架伸缩比 support extension ratio
带压移架 sliding advance of support
本架控制 local control
邻架控制 adjacent control
顺序控制 sequential control
成组控制 batch control, bank control
电液控制 electro-hydraulic control
主顶梁 main canopy
前探梁 cantilever roof bar
伸缩(前)梁 extensible canopy
掩护梁 caving shield
护帮板 face guard
底座 base
双纽线机构 lemniscate linkage
乳化液泵站 emulsion power pack
防倒装置 tilting prevention device
防滑装置 slippage prevention device

Unit 7　Mine Ventilation

矿 井 通 风

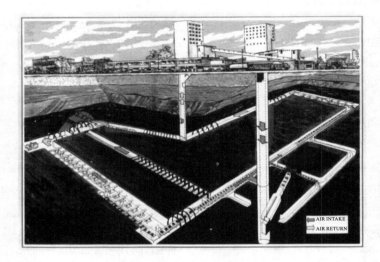

Fig. 7.1　Schematic diagram of mine ventilation roadways

Text 1　Importance of Mine Ventilation Systems
矿井通风系统的必要性

　　Ventilation is sometimes described as the lifeblood of a mine, the intake airways being arteries that carry oxygen to the working areas and the returns veins that transport pollutants away to be expelled to the outside atmosphere. Without an effective ventilation system, no underground facility that requires personnel to enter it can operate safely.

　　The slaughter of coal miners that took place in the coal mines of Britain during the 18th and 19th centuries resulted in the theory and art of ventilation becoming the primary mining science. The success of research in this area has produced tremendous improvements in underground environment conditions.

Unit 7 Mine Ventilation 矿井通风

Loss of life attributable to inadequate ventilation is now, thankfully, relatively infrequent occurrence. Improvements in ventilation have also allowed the productivity of mines to be greatly improved. Neither the very first nor the very latest powered machines could have been introduced underground without an adequate supply of air. Subsurface ventilation engineers are caught up into an odd cyclic process. Their work allows rock to be broken in ever larger quantities and at greater depths. This, in turn, produces more dust, gases and heat, resulting in a demand for yet better environmental control.

1. The Objectives of Subsurface Ventilation

The basic objective of an underground ventilation system is clear and simple. It is to provide airflows in sufficient quantity and quality to dilute contaminants to safe concentrations in all parts of the facility where personnel are required to work or travel. This basic requirement is incorporated into mining law in those countries that have such legislation. The manner in which "quantity and quality" are defined varies from country to country depending upon their mining history, the pollutants of greatest concern, the perceived dangers associated with those hazards and the political and social structure of the country. The overall requirement is that all persons must be able to work and travel within an environment that is safe and which provides reasonable comfort. An interpretation of the latter phrase depends greatly on the geographical location of the mine and the background and expectations of the workforce. Personnel in a permafrost mine work in conditions that would be unacceptable to miners from an equatorial region, and vice versa, and neither set of conditions would be tolerated by factory or office workers. This perception of "reasonable comfort" sometimes causes misunderstandings between subsurface ventilation engineers and those associated with the heating and ventilating industry for buildings.

While guaranteeing the essential objectives related to safety and health, subsurface environmental engineering has, increasingly, developed a wider purpose. In some circumstances, atmospheric pressure and temperature may be allowed to exceed the ranges that are acceptable for human tolerance. For example, in an underground repository for high level nuclear waste, a containment drift may be sealed against personnel access after emplacement of the waste canisters has been completed. However, the environment within the drift must still be maintained to such extent that rock wall temperatures are controlled. This is necessary to enable the drift to be reopened relatively quickly for retrieval of the nuclear waste at any subsequent time during the active life of the repository. Other forms of underground storage often require environmental control of pressure, temperature and humidity for the preservation of the stored material. Yet another trend is towards automated (manless) working faces and the possible use of underground space for in-situ mineral processing. In such zones of future mines, environmental control will be required for the efficient operation of machines and processes, but not necessarily with an atmosphere acceptable to the unprotected human physiology.

2. Factors that Affect the Underground Environment

During the development and operation of a mine or other underground facility, potential hazards arise from dust, gas emissions, heat and humidity, fires, explosions and radiation. Table 7.1 shows the factors that may contribute towards those hazards. Its divisions are based upon natural characteristics and design decisions on how to develop and operate the facility.

Table 7.1 Factors in the creation and control of hazards in the subsurface environment

Factors that contribute to			Methods of control	
NATURAL FACTORS	DESIGN FACTORS	HAZARD	ANCILLARY CONTROL	AIRFLOW CONTROL
Depth below surface	Method of mining	Dust	Dust suppression	Main fans
Surface climate	Layout of mine or facility			
Geology	Rate of rock fragmentation	Gas emissions	Gas drainage	Booster fans, auxiliary ventilation
Physical and chemical properties of rock	Mineral clearance			Natural ventilation
	Method of working	Heat and humidity	Refrigeration systems	
Gas content of strata	Type, size and siting of equipment	Fires and explosions		Airlocks, stoppings, air crossings, regulators
Subsurface liquids	Vehicular traffic		Monitoring systems	
Age of openings	Stored materials	Radiation		Number, size and layout of openings

The major method of controlling atmospheric conditions in the subsurface is by airflow. This is produced, primarily, by main fans that are usually, but not necessarily, located on surface. National or state mining law may insist that main fans are sited on surface for gassy mines. While the main fan, or combination of main fans, handles all of the air that circulates through the underground network of airways, underground booster fans serve specific districts only. Auxiliary fans are used to pass air through ducts to ventilate blind headings. The distribution of airflow may further be controlled by ventilation doors, stoppings, air crossings and regulators.

It is often the case that it becomes impracticable or impossible to contend with all environmental hazards by ventilation alone. For example, increases in air temperature caused by compression of the air in the downcast shafts of deep mines may result in that air being too hot for personnel even before it comes into operation. No practical amount of increased airflow will solve that problem. Table 7.1 includes the ancillary control measures that may be advisable or necessary to supplement the ventilation system in order to maintain acceptable conditions underground.

3. The Integration of Ventilation Planning into Overall System Design

The design of a major underground ventilation and environmental control system is a complex process with many interacting features. The principles of systems analyses should be applied to ensure that the consequences of such interaction are not overlooked. However, ventilation and the underground environment should not be treated in isolation during planning exercises. They themselves, are an integral part of the overall design of the mine or subsurface facility.

It has often been the case that the types, numbers and sizes of machines, the required rate of mineral production and questions of ground stability have dictated the layout of a mine without, initially, taking the demands of ventilation into account. This will result in a ventilation system that may lack effectiveness and, at best, will be more expensive in both operating and capital costs than would otherwise have been the case. A common error is related to size shafts that are appropriate for the hoisting duties but inadequate for the long term ventilation requirement of the mine. Another problem in point is a ventilation infrastructure that is adequate for an initial layout but lacks the flexibility to handle fluctuating market demands for the mineral. Again, this can be very expensive to correct. The results of inadequate ventilation planning and system design are premature cessation of production, high costs of reconstruction, poor environmental conditions and, still too often, tragic consequences to the health and safety of the workforce. It is, therefore, most important that ventilation engineers should be incorporated as an integral part of a design team from the initial stages of planning a new mine or other underground facility.

Text 2　Mine Ventilation Systems　矿井通风系统

Fig. 7.2 depicts the essential elements of a ventilation system in an underground mine or other subsurface facility.

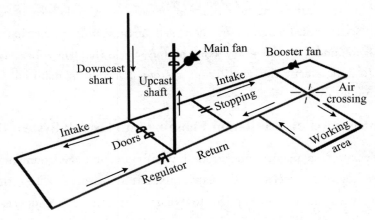

Fig. 7.2　Typical elements of a main ventilation system

Fresh air enters the system through one or more downcast shafts, drifts (slopes, adits) or other connections to surface. The air flows along intake airways to the working areas or places where the majority of pollutants are added to the air. These include dust and a combination of many other potential hazards including toxic or flammable gases, heat, humidity, and radiation. The contaminated air passes back through the system along return airways. In most cases, the concentration of contaminants is not allowed to exceed mandatory threshold limits imposed by law and safe for the entry of personnel into all parts of the ventilation system including return airways. The intake and return airways are often called intakes and returns for short respectively. The return air eventually passes back to the surface via one or more upcast shafts, or through

Unit 7　Mine Ventilation　矿井通风

inclined or level drifts.

1. Fans

The primary means of producing and controlling the airflow are also illustrated on Fig. 7.2 Main fan, either singly or in combination, handle all of the air that passes through the entire system. These are usually, but not necessarily, located on surface, either exhausting air through the system as shown of Fig. 7.2 or, alternatively, connected to downcast shafts or main intakes and forcing air into and through the system. Because of the additional hazards of gases and dust that may both be explosive, legislation governing the ventilation of coal mines is stricter than that of most other underground facilities. In many countries, the main ventilation fans for coal mines are required, by law, to be placed on surface and may also be subject to other restrictions such as being located out of line with the connected shaft or drift and equipped with "blow-out" panels to help protect the fan in case of a mine explosion.

2. Stoppings and Seals

In developing a mine, connections are necessarily made between intakes and returns. When these are no longer required for access or ventilation, they should be blocked by stoppings in order to prevent short-circuiting of the airflow. Stoppings can be constructed from masonry, concrete blocks or fireproofed timber blocks. Prefabricated steel stoppings may also be employed. They should be well keyed into the roof, floor and sides, particularly if the strata are weak or in coal mines liable to spontaneous combustion. Leakage can be reduced by coating the high pressure face of the stopping with a sealant material and particular attention paid to the perimeter. Here again, in weak or chemically active strata, such coatings may be extended to the rock surfaces for a few meters back from the stopping. In case where the airways are liable to convergence, precautions should be taken to protect stoppings against premature failure or cracking. These measures can vary from "crush pads" located at the top of the stopping to sliding or deformable panels on prefabricated stoppings. In all cases, components of stoppings should be fireproof and should not produce toxic fumes when heated.

As a short term measure, fire-resistant brattice curtains may be tacked to roof, sides and floor to provide temporary stoppings where pressure differentials are low such as in locations close to the working areas.

Once abandoned areas of a mine are to be isolated from the current ventilation infrastructure, seals should be constructed at the entrances of the connecting airways. If required to be explosion-proof, these consist of two or more stoppings, 5 m to 10 m apart, with the intervening space occupied by sand, stone dust, compacted non-flammable rock waste, cement-based fill or other manufactured material. Steel girders, laced between roof and floor add structural strength. Grouting the surrounding strata adds to the integrity of the seal in weak ground. In coal mines, mining law or prudence in safety may require seals to be explosion-proof.

3. Doors and Airlocks

When access must remain available between an intake and a return airway, a stopping may be fitted with a ventilation door. In its simplest form, this is merely a wooden or steel door hinged such that it opens towards the higher air pressure. This self-closing feature is supplemented by angling the hinges so that the door lifts slightly when opened and closes under its own weight. It is also advisable to fit doors with latches to prevent their opening in cases of emergency when the direction of pressure differentials may be reversed. Contoured flexible strips attached along the bottom of the door assist in reducing leakage, particularly when the airway is fitted with rail track.

Ventilation doors located between main intakes and returns are usually built as a set of two or more to form an airlock. This prevents short-circuiting when one door is opened for passage of vehicles or personnel. The distance between doors should be capable of accommodating the longest train of vehicles required to pass through the airlock. For higher pressure differentials, multiple doors also allow the pressure break to be shared between doors.

Mechanized doors, opened by pneumatic orelectrical means are particularly convenient for the passage of vehicular traffic or where the size of the door or air pressure would make manual operation difficult. Mechanically operated doors may, again, be side-hinged or take the form of roll-up or concertina devices. They may be activated manually by a pull-rope or automatic sensing of an approaching vehicle or person. Large doors may be fitted with smaller hinged openings for access by personnel. In such cases, a sliding panel may be fitted in order to reduce that pressure differential temporarily while the door is opened. Interlock devices can also be employed on an airlock to prevent all doors from being opened simultaneously.

4. Regulators

A passive regulator is simply a door fitted with one or more adjustable orifices. Its purpose is to reduce the airflow to a desired value in a given airway or section of the mine. The most elementary passive regulator is a rectangular orifice cut in the door and partially closed by a sliding panel. The airflow may be modified by adjusting the position of the sliding panel manually. Louvre regulators can also be employed. Another form of regulator is a rigid duct passing through an airlock. This may be fitted with a damper, louvers or butterfly valve to provide a passive regulator or a fan may be located within the duct to produce an active regulator. Passive regulators may be actuated by motors, either to facilitate their manual adjustment or to react automatically to the detected changes in the quantity or quality of any given airflow.

When the airflow in a section of the mine must be increased to a magnitude beyond that obtainable from the system then this may be achieved by active regulation. This implies the use of a booster fan to enhance the airflow through that part of the mine. Once booster fans are employed, they should be designed into the system such that they help control leakage without causing undesired recirculation in either normal or emergency situations. However, in some countries, coal mine legislation prohibits the use of booster fans.

5. Air Crossings

When intake and return airways are required to cross over each other than leakage between the two must be controlled by the use of an air crossing. The sturdiest form is a natural air crossing in which the horizon of one of the airways is elevated above the other to leave a sill of strata between the two, perhaps reinforced by roof bolts, girders or timber boards. A more usual method is to intersect the two airways during construction, then to heighten the roof of one of them or excavate additional material from the floor of the other. The two airstreams can then be separated by horizontal girders and concrete blocks, or a steel structure with metal or timber shuttering. Sealants may be applied on the high pressure side. Control of the airway gradients approaching the air crossing reduces the shock losses caused by any sudden change of airflow direction. Man-doors can be fitted into the air-crossing for access.

Completely fabricated air crossings may be purchased or manufactured locally. These can take the form of a stiffened metal tunnel. Such devices may offer high resistance to airflow and should be sized for the flow they are required to pass. They are often employed for conveyor crossings. Another type of air crossing used mainly for lower airflows and requiring no additional excavation is to course one of the airstreams through one or more ducts that intersect a stopping on either side of the junction. An advantage of this technique is that the ducted airflow may be further restricted by passive regulators or up-rated by fans in the ducts.

In all cases, the materials used in the construction of air crossings should be fireproof and capable of maintaining their integrity in case of fire. Neither aluminum nor any other low melting-point or combustible material should be employed in an air crossing.

Depending on the type of mine and disposition of local geology, ventilation layouts can be divided into two broad classifications: either a U-tube system or a through-flow arrangement (Fig. 7.3). Fig. 7.3(a) shows a basic U-tube configuration where air flows towards and through the working area, then returns along adjacent airways, often separated from intakes by long pillars and/or stoppings. Access doors in the stoppings facilitate traffic between intake and return airways. The variation of this arrangement would be room-and-pillar and longwall type mining methods. The other arrangement is shown in Fig. 7.3(b), where intakes and returns are usually separated geographically from adjacent airways, which are either all intakes or all returns. Although fewer stoppings and airways are needed because of the geographical separation, which often results in less air leakage, air current regulations and boosters may be required for airflow control in work areas. Parallel flows between intake and return air shafts across the multilevel metal mines and the bleeder system in a longwall panel would be typical examples of this type layout.

Two factors that have significantly bearing on the design of the longwall ventilation systems are the control of methane or other gases that accumulate in the gob area and the increasing high rate of rock breakage on heavily mechanized longwalls that has exacerbated the production of dust, gas, heat, and humidity. Fig. 7.4 depicts some of the commonly used ventilation layouts used on longwall sections. In the USA, a minimum of two entries is required, while single entry longwalls are primarily employed in European coal mines.

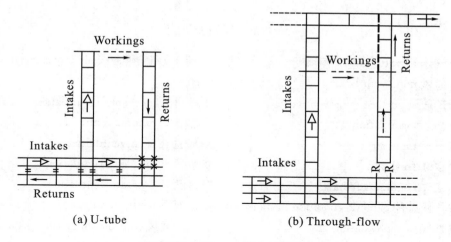

Fig. 7.3 Basic ventilation systems

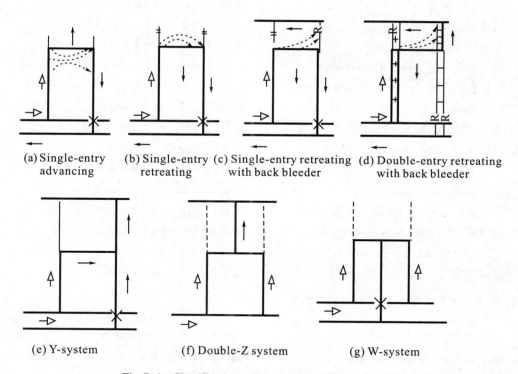

Fig. 7.4 Classifications of longwall ventilation systems

System layouts become more complex when mining under inclined, thick, and gassy coal seams with frequent faults. Narrower and shorter panels are necessary to cope with these difficult conditions. Futhermore, other type of layouts are also ready-to-use to accommodate specific geological conditions.

Terms 常见表达

有害气体 harmful gas
矿井通风 mine ventilation
露天矿通风 surface mine ventilation, open-pit ventilation
矿井空气 mine air
新鲜空气 fresh air
污浊空气 contaminated air
矿井气候条件 climatic condition in mine
窒息性气体 asphyxiating gas
可燃性气体 inflammable gas
矿井空气调节 mine air condition
风量 air quantity
需风量 required airflow
风量按需分配 air distribution
风量自然分配 natural distribution of air flow
进风风流 intake air flow
回风风流 return air flow
负压 negative pressure
正压 positive pressure
自然风压 natural ventilation pressure
通风机全压 total pressure of fan
通风机动压 velocity pressure of fan
通风机静压 static pressure of fan
通风压力分布图 ventilation pressure map
通风阻力 ventilation resistance
摩擦阻力 frictional resistance
局部阻力 shock resistance
摩擦阻力系数 coefficient of frictional resistance
局部阻力系数 coefficient of shock resistance
等积孔 equivalent orifice
(井巷)风阻特性曲线 air way characteristic curve
机械通风 mechanical ventilation
自然通风 natural ventilation
局部通风 local ventilation
扩散通风 diffusion ventilation
分区通风 separate ventilation
串联通风 series ventilation
压入式通风 forced ventilation
抽出式通风 exhaust ventilation
上行通风 ascension ventilation
下行通风 declensional ventilation
独立风流 separate airflow
循环风 recirculating air
中央式通风 centralized ventilation
对角式通风 radial ventilation
混合式通风 compound ventilation
串联网络 series network
并联网络 parallel network
角联网络 diagonal network
通风网络图 ventilation network map
通风系统图 ventilation map
主要通风机 main fan
局部通风机 auxiliary fan
辅助通风机 booster fan
水力引射器 water jet fan
通风机效率 fan efficiency
通风机特性曲线 fan characteristic curve
通风机工况点 fan operating point
通风机全压输出功率 fan total air power, total output power of fan
通风机静压输出功率 fan static power, static output power of fan
通风机全压效率 fan total efficiency

Unit 7 Mine Ventilation 矿井通风

通风机静压效率 fan static efficiency
通风机附属装置 accessory equipment of fan
风硐 fan drift
扩散器 fan diffuser
防爆 breakaway explosion door
反风道 air-reversing way
风量调节 air regulation
矿井有效风 effective air quantity
漏风 air leakage
矿井外部漏风 surface leakage
矿井内部漏风 underground leakage
矿井内部漏风率 underground leakage rate
矿井外部漏风率 surface leakage rate
矿井有效风量率 ventilation efficiency
风桥 air crossing
风门 air door
反风 air reversing
风窗 air regulator
风墙 air stopping
风障 air brattice
风筒 air duct
测风站 air measuring station
风表校正曲线 calibration curve of anemometer
检定管 detector tube
粉尘采样器 dust sampler
甲烷报警器 methane alarm
瓦斯测定器 gas detector
甲烷遥测仪 remote methane-monitor
甲烷断电仪 methane monitor breaker
风电闭锁装置 fan-stoppage breaker
风电甲烷闭锁装置 fan-stoppage methane-monitor breaker
电化学式瓦斯测定器 electrochemical type gas detector
粉尘 dust
矿尘 mine dust
煤尘 coal dust
岩尘 rock dust
浮尘 airborne dust
落尘 settled dust
呼吸性粉尘 respirable dust
粉尘粒度分布 dust size distribution
游离二氧化硅 free silica
隔爆 explosion suppression
粉尘浓度 dust concentration
煤尘爆炸 coal explosion
煤尘爆炸危险煤层 coal seam liable to dust explosion
尘肺病 pneumoconiosis
硅肺病 silicosis
煤肺病 anthracosis
煤硅肺病 anthraco-silicosis
煤矿防尘 dust suppression in mine
煤层注水 coal seam water infusion
水幕 water curtain
矿井火灾 mine fire
水棚 barrier with water
岩粉 rock dust
岩粉棚 rock dust barrier
自动隔爆装置 triggered barrier
防尘口罩 dust mask

Unit 8   Mine Auxiliary Production System

矿井辅助生产系统

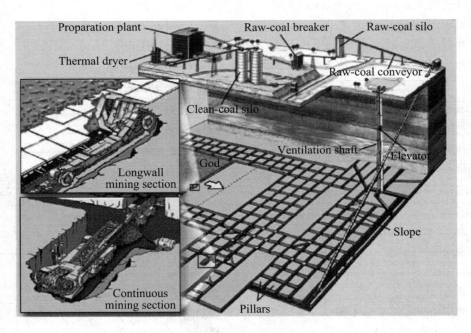

Fig. 8.1   Schematic diagram of mine auxiliary production system

To extract coal from underground coal seam safely and effectively, several systems are needed, such as mine haulage system, mine hoisting system, drainage system, radio communication and monitoring system and so on. The mine development system is an important part of the mining technology of a coal mine, and it influences directly the producing efficiency and benefit. Along with the progress of science and technology, the coal mine development system has to reform and to get perfect unceasingly, so as to guarantee the expand of mine production and the establishment of modernized coal mines.

Unit 8　Mine Auxiliary Production System　矿井辅助生产系统

Text 1　Mine Haulage System　矿井运输系统

Since mechanization in coal mines began with haulage, the history of mine haulage is, basically, the history of coal mine mechanization and, especially, the evolution of mine power systems.

In the late 1700s, men were employed for mine haulage and human power gave way progressively to dogs, ponies, mules, horses, and finally mechanized power, although the period of transformation was a long one.

While trolley locomotive track or rail haulage was the initial form of mechanized haulage, and still remains important, many other forms of vehicular motive power exist with more changes portended for the future. Conveyor haulage has become increasingly important, and we appear on the threshold of massive utilization of hydraulic transportation.

Locomotive rail haulage was first introduced in this country in an underground anthracite coal mine in the 1870s, and, to this day, rail haulage in this country has been evolved electrically power trolley. Locomotive haulage was responsible for the introduction of D-C power systems into USA coal mines since a D-C system is the simplest electrical arrangement, requiring only two conductors, one of which can be the rail; also the D-C motor possesses the best haulage characteristics. Because of the presence of the D-C system for haulage purposes, other mine equipment components were added to it as they became available until about 1950 when the mass introduction of continuous mining overloaded the D-C system and the swing began to A-C. Although few mines with a D-C primary power system will be found today, D-C power is used almost exclusively with rail haulage and also widely for shuttle cars. Virtually all the motors of large stripping equipment are D-C powered.

While compressed-air locomotive haulage was formerly practiced to some degree, it is obsolete today. Battery locomotives are a possible supply-manpower alternative that are rarely employed for coal haulage, although batteries are used for trackless equipment that hauls coal as well as supplies and personnel. The diesel engine was invented in 1882, but its practical application was not demonstrated until 1897. In 1927, a diesel engine was used for the first time in underground haulage equipment in a German coalmine. The first use of diesel-powered equipment in underground mines in this country is believed to have been in a Pennsylvama limestone mine in 1939. Introduction of diesel trucks and other equipment into many mines and tunnels in different areas followed shortly thereafter.

In general, main-line transportation differs from intermediate transportation only in size, scope, and permanence of installation except for mines in which hydraulic systems are used or small mines in which battery-powered equipment hauls from the face to the surface. The fundamental difference between face transportation and other transportation should be appreciated. Face transportation is normally the major in production, so that the object in its section or design is simply to build as much capacity as possible into it. Intermediate and main haulage, which may be considered as service functions to face haulage, present prominent problems in providing adequate capacity in the most economic manner.

Text 2   Mine Hoisting System   矿井提升系统

A shaft provides access to the network of openings used to recover the underground resource, serves as an escape way in case of emergency, and allows vertical movement of miners and materials, with the help of mine hoisting system.

It is difficult to understate the importance of having in any underground mine serviced by a vertical shaft, a hoisting system that is both safe and productive. Fortunately, experience indicates that safety and productivity mutually reinforcing each other, you cannot have one without the other and therefore by corollary, if the hoisting system is designed to be very safe, and operated and maintained that way, then it will almost certainly be highly productive as well.

In recent years, the drive towards improved efficiency and productivity has led to many technical changes in the design, operation and maintenance of mine hoisting systems. In particular, in Australia, new hoisting systems are highly automated and, with the exception of some maintenance and testing procedures, operating totally unattended, that is: without a driver, operator, platman or onsetter. Supervision of the hoisting system is normally conducted from a remote, central mine-monitoring facility.

The earliest unattended, automatic hoisting systems were installed in Australia over twenty years ago and were characteristic of quite intensive

Unit 8 Mine Auxiliary Production System 矿井辅助生产系统

maintenance. However, provided maintenance was of the highest standard, all the evidence suggests that these hoisting systems can provide a very high degree of operational safety as well as improved performance.

The maintenance of hoisting systems has also changed dramatically. Routine maintenance is reduced to a minimum but condition monitoring and preventative maintenance is embraced. Testing, particularly statutory testing, is increasingly time consuming; hence the current emphasis is put on automated testing procedures and recording. The implementation of these facilities requires careful consideration during the system design phase.

Usually, a complete mine hoisting system includes:

1. Hoist Brake System

This system controls braking works in the following two ways: Braking during abnormal operation with proportional control to give a soft stop; Braking at an emergency stop with control giving a constant predefined retardation independent of the braking conditions. The system includes battery backup to provide correct operation during a power failure as well as monitoring of tachometers, supply voltages and hydraulic pressure. In the event of a battery failure, the hoist will stop within a safe retardation rate.

2. Control and Monitoring System

Installation of a new control system or the upgrade of an existing one leads to increased safety, increased productivity and increased availability of the hoisting plant. The control system solutions should be the complete solution for safety and monitoring of all mine hoist types, which provides very accurate monitoring of all vital hoist parameters such as speed, acceleration, retardation and position of the hoist conveyances.

A modern control system will result in:
- Reduced hoisting cycle time and increased production.
- Improved operator functionality.
- Improved reporting for production and maintenance.
- Advanced diagnostics for fault tracing.
- Possibility for remote diagnostics.

3. Hoist Drive Systems

Coal mine hoists are usually powered by reliable A-C or D-C motor drive systems, operating either directly or through a gearbox or flexible coupling. In order to minimize stress in the ropes the drive systems should be designed to go on smooth changes in motor torque and speed.

Text 3　Mine Drainage System　矿井排水系统

Drainage in coal mines is extremely variable. In some mines, the drainage is excellent and water causes few problems; in others, lack of drainage results in severe inundations. However, even a small amount of water in low coal can bring discomfort to miner and result in reduced productivity.

There are both direct and indirect problems associated with the influx of water into a coal mine; these have effects in as well as out of the mine. Some of the direct effects of water in a mine might include:

- The blocking of mine entries to the passage of air and haulage by the accumulation of water.
- Interruption of production and damage to the mine, possibly even loss of life due to inrushes of water.
- Increased costs caused by the need to remove the water.
- Interference with haulage; especially, water can cause poor traction for rubber-tired equipment and result in track deterioration as it washes away ballast.

To minimize the problems associated with an influx of unwanted water into underground mines, a four-step process of control is suggested as follows: prevention, collection, transportation, and disposal. Each of these will be reviewed briefly.

1. Prevention

Each gallon of water that is prevented from entering the mine one less gallon that will have to be collected, transported, and disposed of. Some common sense precautions such as avoiding the sitting of shafts, boreholes, and other opening in the slow spots on the surface and other measures can prevent surface runoff from entering the mine.

Unit 8　Mine Auxiliary Production System　矿井辅助生产系统

2. Collection

Even with the best mine design and all sorts of precautionary measures, water influx into mines can not be entirely eliminated. The water that accumulates has to be collected before it can be transported and disposed of. In the collection process, it is imperative that gravity be used as much as possible to minimize the power required to transport the water.

An interesting collection system for mine water is the use of the diversion tunnel. The concept is to predrain a mining area by locating a drainage tunnel under the seams to be mined. If the water can be drained from an area prior to mining, subsequent mining may be done without any water problems. This method appears very attractive since not only does it preclude the handling of water in mines but also the water will remain pure and uncontaminated.

3. Transportation

Even where gravity has been utilized to the greatest possible degree for collecting mine water, some means of transporting the water out of the mine is still required. Today, pumps are used almost exclusively for this purpose.

4. Disposal

With the high effluent disposal standards that must be maintained for streams today, disposal of mine water can be a very costly problem. There must be some knowledge of the material to be disposed of if it is to be proposed when it comes to proper handing.

Oxidation of pyrites in the coal seam and strata overlying and underlying the seam is the initial step in the formation of acid mine water. As oxidation continues, the material disintegrates, exposing new surfaces for further oxidation and acid formation. Thus time is an important factor: the longer the acid forming materials are exposed to the atmosphere, the greater the amount of acid that will be formed. Thus the priority for treating mine water is designed to neutralize acidity and remove iron by processes involving the use of lime or limestone and by demineralization.

Mine drainage systems contain such items as sumps, suction pipes, pumps, discharge pipes, and appropriate fittings such as valves and ells, Fig. 8. 2, to move water from a point in the mine to the surface. Energy is introduced into the system via the pump to overcome the following total dynamic water heads (H).

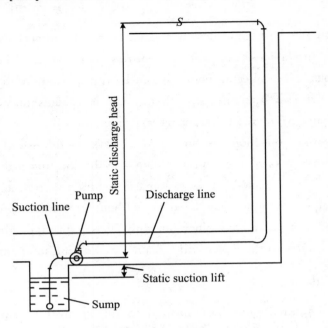

Fig. 8.2 Mine drainage system

Terms 常见表达

矿井建设 mine construction
井巷工程 shaft sinking and drifting
矿山地面建筑工程 mine surface construction engineering
矿场建筑工程 mine plant construction engineering
安装工程 installation engineering
斜井 inclined shaft
平硐 adit
主平硐 main adit
隧道 tunnel
井筒 shaft, slant
井口 shaft mouth
井颈 shaft collar
井身 shaft body
井窝 shaft sump

主井 main shaft
副井 auxiliary shaft, subsidiary shaft
箕斗井 skip shaft
罐笼井 cage shaft
混合井 skip-cage combination shaft
风井 ventilating shaft, air shaft
矸石井 waste shaft
马头门 ingate
井底车场 pit bottom, shaft bottom
环行式井底车场 loop-type pit bottom
折返式井底车场 zigzag-type pit bottom
硐室 (underground) room, (underground) chamber
井底煤仓 shaft coal (-loading) pocket
翻车机硐室 tippler pocket, rotary dump room

Unit 8　Mine Auxiliary Production System　矿井辅助生产系统

箕斗装载硐室 skip loading pocket
主排水泵硐室 main pumping room
水仓(drain) sump
吸水井 draw-well, suction well
配水巷 water distribution drift
井下充电室 underground battery-charging station
井下主变电硐室 underground central substation room
井下(机车)修理间 underground locomotive repair room
井下调度 underground control room, pit-bottom dispatching room
(井下)等候室 pit-bottom waiting room
爆炸材料库 magazine
躲避硐 manhole, refuge pocket
腰泵房 stage pump room
巷道 drift, roadway
水平巷道(horizontal)drift, entry
倾斜巷道 inclined drift
岩石巷道 rock drift
煤巷 coal drift
煤-岩巷 coal-rock drift
人行道 pedestrain way, sidewalk
交岔点 intersection, junction
井筒安全道 escapeway
管子道 pipe way
暖风道 preheated-air inlet
检修道 maintainin groadway
单项工程 individual project
单位工程 unit project
矿井建设周期 mine construction period
矿井施工准备期 preparation stage of mine construction
井巷过渡期 transition stageshaft to drift
建井期 mine construction stage
矿井建设总工期 overall stage of mine construction
矿井建设关键线路 critical path of mine construction

Unit 9 Mine Subsidence and Control

开采沉陷与控制

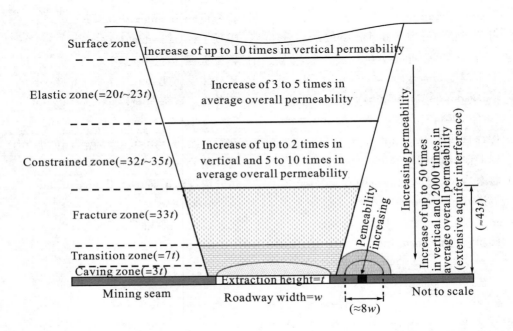

Fig. 9.1 Diagram of mine subsidence zones of underground coal mining

Text 1 Introduction to Mining Subsidence 开采沉陷概述

1. The General Meaning of Subsidence

The term "subsidence" commonly refers to a point (or points) on or in the earth's surface moving vertically to a lower level. To understand and manage any potential impacts of subsidence, it is important to pay attention to both the vertical and horizontal component of these movements.

Subsidence can result from natural causes such as earthquakes and from human activities such as removal of materials from below the surface or extraction of groundwater from alluvial aquifers, see Fig. 9.2.

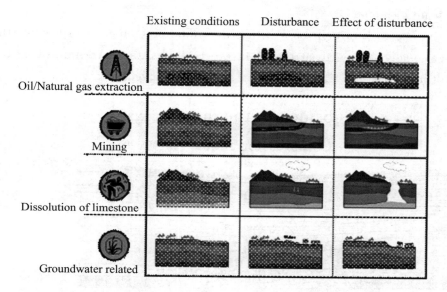

Fig. 9.2 A graphic illustrating the different types of ground subsidence

2. Mine Subsidence

Mine subsidence can be defined as movement of the ground surface as a result of readjustments of the overburden due to collapse or failure of underground mine workings. Surface subsidence features usually take the form of either sinkholes or troughs. The extent of areas affected by subsidence due to underground coal mining can be more significant than subsidence resulting from other forms of activity such as tunneling. Therefore this text specifically refers to "mine subsidence" as being subsidence caused by underground coal mining.

"Longwall" mining is the safest, most efficient and most commonly used method of underground coal mining in Australia and other developed countries. Although this method is the major cause of mine subsidence in NSW (New South Wales), it should not be viewed as the sole cause of mine subsidence. To help understand "mine subsidence", Fig. 9.3 shows a vivid illustration:

• The extraction of coal may cause subsidence of the rock and soil mass above the extracted coal seam. The area of subsidence influence is shown in Fig. 9.3.

• The limit, magnitude and nature of subsidence movements are dependent on numerous factors, such as mine design, geological conditions, surface topography, and the distance between the mine workings and the ground surface (or point of interest).

• The rock or soil mass within the influence of mining may move three-dimensionally at any given point. In other words, the extraction of coal may result in vertical as well as horizontal subsidence movements.

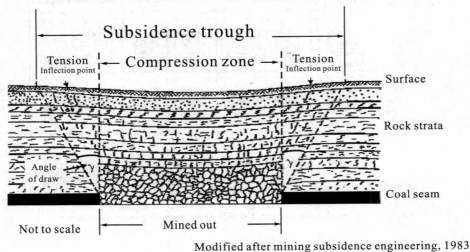

Modified after mining subsidence engineering, 1983

Fig. 9.3 Diagram of mine subsidence development from underground coal mining

3. Understanding the Potential Impacts of Mine Subsidence

The development of mine subsidence can potentially impact on both natural and man-made surface and sub-surface features. The management of these impacts requires an understanding of fundamental aspects of mine subsidence such as:

• The magnitude and direction of subsidence movements at any point within the influence of mining.

• Differential subsidence movements between parts of a given surface or sub-surface feature. Such differences (either in the direction or magnitude of movements) may lead to impacts on a surface or sub-surface feature. For example, if every part of a building (or a given natural feature) is subject to movement of the same magnitude in one direction, this building is unlikely to be damaged.

• The tolerance of a given surface or sub-surface feature to the influence of mine subsidence. Depending on their characteristics and site-specific conditions, the natural environment and man-made structures can withstand certain levels of mine subsidence. Not every feature will be affected. The occurrence, nature, severity, duration or "reparability" of subsidence impacts depend strongly on the magnitude or nature of subsidence, and the nature of the environmental system or man-made structure in question.

4. Subsidence Prediction and Management of Impacts

The prediction and management of mine subsidence aims to produce outcomes that are consistent with community and government expectations for responsible mining. The required management outcomes cannot be achieved without adequate understanding and assessment of site-specific factors that may affect the development of subsidence, and the responses of the environmental system or man-made structures in question.

Like many other engineering disciplines involving geological materials, uncertainties are an inherent part of subsidence engineering and management. Management of subsidence must be risk-based, flexible, responsive and capable of dealing with unexpected changes or uncertainties.

In recent years, significant progress has been made towards achieving a better understanding of subsidence development related to the natural environment. This has led to improvements in the accuracy of mine subsidence impact predictions.

The NSW government actively encourages the use of new mining techniques which have the potential to minimize subsidence or the effects of subsidence. The mining industry is in the process of developing and advancing corresponding technologies to meet these objectives.

Text 2　Mining Subsidence and Ground Control Technology 开采沉陷与矿压控制技术

Due to a great deal of coal resources being extracted from underground, the environmental hazards resulted from mining activities are becoming a serious problem. Coal mining subsidence not only destroys the ecological environment, but also causes surface structure damage (Fig. 9.4 and Fig. 9.5).

Fig. 9. 4 A road destroyed by subsidence and shear, near Castleton, Derbyshire

Fig. 9. 5 Severe cracking of supporting walls and footings in structures

In order to control mining subsidence and to protect surface structures, water bodies and railway lines from damages, it is necessary to carry out research mining subsidence and ground control technology. Through several decades of studying and practice, China has accumulated abundant experience on mining subsidence control technology.

1. Strip Pillar Mining Method

Strip pillar mining is the most widely used method to control mining subsidence in China. In strip pillar mining, the coal reserve is divided into regular strips with alternate strips being extracted. The strips left behind, called strip pillars, are designed to support the overburden and prevent surface subsidence. This is one of the important methods in "Green Mining Technology" and has become an effective method to mine those coal reserves lying under village structures.

The advantage of strip pillar mining is to reduce the surface subsidence effectively without changing mining technology. It began to be put into use in 1976, and currently a large amount of coal reserves under "3-body" have been extracted by the strip pillar mining method. In China, the roof control method of strip pillar mining is almost by caving method. The mining depth of strip pillar mining is less than 500 m; mining height is mostly less than 6 m, and the recovery ratio ranges between 40% and 68%. The surface subsidence factor depending on the recovery ratio is mostly less than 0.2 (Table 9.1).

Table 9.1 Relationship between surface subsidence factor and recovery ratio of strip pillar mining

recovery/%	hard stratum	medium-hard stratum	soft stratum
60	0.09~0.11	0.13~0.17	0.17~0.21
50	0.05~0.06	0.08~0.10	0.10~0.12
40	0.026~0.032	0.03~0.05	0.05~0.06

2. Room and Pillar Method

The room and pillar method is widely used in America, Australia, Canada, India, and South Africa. Its subsidence factor ranges between 0.35 and 0.68. This method has also been employed in recent years in China, but on a small scale. Huangling Coalmine of Shanxi province is the first coal mine to apply room and pillar mining using continuous miners. Other mines with room and pillar mining include those at Nantun Coalmine of Yanzhou Mining Group and Daliuta mine of Dongsheng mining district.

3. Backfilling Mining Methods

(1) Traditional backfilling method

The traditional backfilling method includes hydraulic backfilling, pneumatic backfilling, mechanical backfilling, and coal gangue sliding backfilling. In the process of mining, filling materials such as sands, coal gangue or fly ashes are filled in the gob behind the working face in order to support the overburden strata. The hydraulic backfilling method is the most effective way to control surface subsidence. Its subsidence factor in general ranges from 0.1 to 0.3. When water supply is lacking or underground working face is damp or watery, the pneumatic backfilling method should be employed. Coal gangue sliding backfilling method is only used in steeply inclined and inclined coal seams; The subsidence factor of the coal gangue sliding filling method is roughly 0.3~0.4.

(2) Paste backfilling method

Backfilling materials and backfilling technology have made great progress in China in recent years. The paste backfilling method is a newly-emerged method in China, and one of the key measures in "green mining technology". This method can increase the coal recovery ratio, protect groundwater and surface structures from damages, improve mining area environment and make use of solid waste materials. Jinchuan Company in China set up the paste backfilling production system in 1996. Now some coalmines in Shandong province are using this method. After paste backfilling materials are filled in the gob, they become non-water-yielding material aggregate, and its solid constituents, usually varies from 76% to 85%. The preparation process of paste filling is as follows: first, materials such as coal gangue, fly ash, industrial slag, bank sand are processed into hydrated paste like toothpaste. Then the toothpaste mixture is transported across underground and pump filled into the gob. Once the paste fill is hardened and set in the gob, the cementation fill body can support overburden strata and control surface subsidence.

4. Grouting of Bed Separations in Overburden Strata

During mining, strata separations occur frequently in the overburden especially under strong and thick strata. Industrial waste materials such as fly ash are injected into the voids of strata separations under high pressure. Grouting of the bed separation caused by mining is achieved by injecting into fractured strata, a mixture of fly ash and industrial waste materials based slurry. The injection is carried out from surface boreholes as shown in Fig. 9.6 and Fig. 9.7. The extent of the slurry injection depends on the structure and characteristics of overburden strata. In general, bed separations in overburden strata are formed under competent and thick strata.

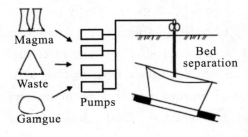
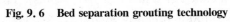
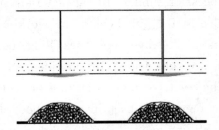

Fig. 9.6　Bed separation grouting technology　　Fig. 9.7　Bed separation grouting under the key strata

In a nutshell, there are various reasons that bed separation grouting can reduce surface subsidence. First, fly ash slurry material has better binding strength. When injected into the bed separation, fly ash can occupy the fractured space and slowly bind the separated layers, thus preventing the overburden strata from subsiding thus minimizing surface subsidence.

China is a big country with abundant coal resources and coal production. But the problems of surface subsidence and environmental hazards due to coal mining are becoming increasingly serious. Therefore it is imperative to have a better understanding of mining subsidence and damage prevention technology.

Terms 常见表达

开采沉陷 mine subsidence
垮落带 caving zone
断裂带 fractured zone
弯曲带 sagging zone
岩层移动 strata movement
地表移动 surface movement
地表移动观测站 observation station of ground movement
地表移动盆地 subsidence basin
移动盆地主断面 major cross-section of subsidence trough
临界变形值 critical deformation value
边界角 limit angle
移动角 angle of critical deformation
裂缝角 angle of break
充分采动角 angle of full subsidence
最大下沉角 angle of maximum subsidence
超前影响角 advance angle of influence
最大下沉速度角 angle of maximum subsidence velocity

影响传播角 propagation angle
拐点偏移距 deviation of inflection point
主要影响半径 major influence radius
主要影响角正切 tangent of major influence angle
充分采动 full subsidence
非充分采动 subcritical extraction
下沉系数 subsidence factor
水平移动系数 displacement factor
采动系数 factor of full extraction
围护带 safety berm
抗变形建筑物 deformation resistant structure
刚性结构措施 rigid structure measure
柔性结构措施 flexible structure measure
滑动层 sliding layer
缓冲沟 buffer trench
导水断裂带 water conducted zone
防水煤岩柱 safety pillar under water-bodies
井筒煤柱开采 shaft pillar extraction

Unit 10 Integrated Coal Mining and Methane Extraction

煤与瓦斯共采

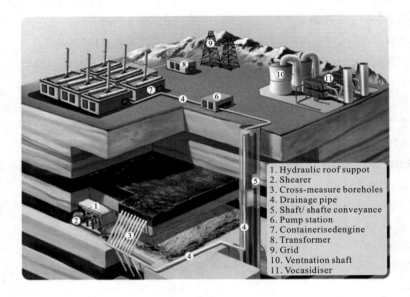

Fig. 10.1 Layout of integrated coal mining and methane extraction

Text 1 Integrated Coal Mining and Methane Extraction
煤与瓦斯共采

Coal is an essential energy resource in China and makes up 74% of China's energy production and consumption. Typically, coal supplies 78% of electricity production, 70% of chemical manufacturing and sixty percent of urban heating and cooking in China. Gas released from coal seams is potentially dangerous during coal mining, not only due to the explosive risk but also the risk of coal and gas outburst. There are many serious gas accidents resulting in numerous

Unit 10   Integrated Coal Mining and Methane Extraction   煤与瓦斯共采

casualties every year in China. 50% of national shafts have high gas content and coal and gas outburst danger. In process of mining, it not only wastes the energy resources but also pollutes the environment seriously when the gas is released directly to the air because the methane has serious greenhouse effect.

1. Basic Idea of Integrated Coal Mining and Methane Extraction

For high-gas coal seams the necessary condition is the safe high-efficient gas extraction if we want to realize the safe high-efficient exploitation of coal. Thus we must eliminate the outburst danger of coal and gas in order to adopt the high-efficient exploitation methods of coal, and we must reduce the quantity of the gas effusing in process of the exploitation of coal in order to make the gas concentration of the returning air not exceed that of the "safety regulation of coal mine". The technical method of realizing these goals is to extract some gas from coal seams and to reduce the gas content of coal seams radically.

The stress distribution of the rock and coal seams above the exploited coal seams not only depends on the exploitation condition (such as the depth of the exploitation, the inclination length of the working face, stoping velocity, physical and mechanical property of the rock seams above the exploited coal seams and so on) but also relys on the key seam which controls the distortion and destruction of the wall rock.

While exploiting coal seams, the coal seam with low gas content free of coal and gas outburst danger was exploited firstly. Then the pressure on the coal seams which are above or below the initial coal seam was relieved making use of the exploiting effect. The distortion and rupture of the coal and rock seams and the crannies will increase the permeability of the coal and rock greatly. So the domino effect of pressure relief, permeability increasing and flowage ability increasing comes into being and the active condition of desorption-diffusion-infiltration flow is formed. The desorption and flowage condition is different because the distribution of crannies is different in various regious. So the high-efficient exploitation of gas can be realized if we adopt reasonable and high-efficient method and system of gas extraction. The extraction of gas reduces the gas content of pressure relief coal seams, eliminates the coal and gas outburst danger of pressure relief coal seams and reduces the quantity of gas effusing to the working face air. Consequently the necessary condition of the safe high-efficient exploitation of coal is created. Fig. 10.2 shows the safe and high-efficient exploitation system of coal and gas in high-gas coal seams.

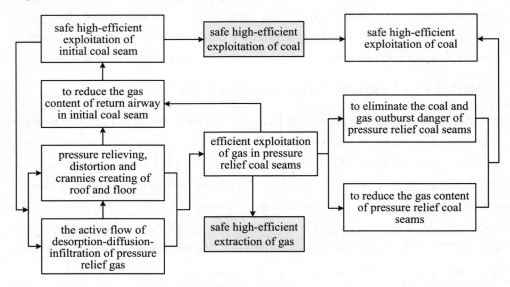

Fig. 10. 2　Safe high-efficient exploitation system of coal and gas in high-gas coal seams

2. Engineering Methods of Integrated Coal Mining and Methane Extraction

The pressure relief degree of the coal seams and the movement condition of the gas differ greatly because of the different comparative intervals after the initial coal seam is exploited. The gas effusion of the gob behind the working face can be divided into short-distance gas emission, intermediate-distance gas emission and long-distance gas emission. The gas from short-distance gas emission is most from the non-exploited separated layers of the initial coal seam, the lost coal in the gob, the coal seams in the caving zone, the lower coal seams of which floor heave greatly and some of the coal seams that are in the fractured zone. The gas of intermediate-distance gas emission is most from the fractured zone and some of the coal seams in the sagging zone. The gas of long-distance gas emission is most from the sagging zone.

(1) Extraction Methods of Short-distance Gas

The successful extraction methods of short-distance gas lie in the strike boreholes in roof through coal and rock seams and the method of strike gas extraction drift in roof. The normal distance from the end of the strike boreholes in roof through coal and rock seams to the roof of the coal seam was not more than 20 m in the mining area of Huainan. 5 or 6 boreholes were constructed in every drill site. The length of the boreholes was 100 m～150 m. The strike horizontal overlapping distance of boreholes in adjacent drill site was 30 m～40 m. The horizontal distance from the acclivitous controlling area of the borehole to the upper air way was 20 m. The best place of gas extraction was 6 m～13 m above the roof of the coal seam. This method played an important role in controlling the short-distance gas emission in different working face of Huainan mining area. The gas flux of single extraction borehole was 1 m^3/min～2 m^3/min. The method of strike gas extraction drift in roof in Huainan mining area was by tunneling a horizontal rock roadway of which cross-section was 3 m^2～4 m^2, which was 20 m away from the upper wind roadway and 10 m～14 m at normal direction away from the roof of the coal seam. Here gas was extracted in the caving arch which was above the gob making use of the roadway.

(2) Extraction Methods of Intermediate-distance Gas

The successful method of intermediate-distance gas extraction lies in strike gas drainage drift in roof in Yangquan mining area. The No. 15 coal seam was exploited in Yangquan mining area. Its average exploitation height was 6.0 m. A horizontal rock roadway was tunneled along the strike of the rock seam to drain the intermediate-distance gas. The cross-section area of the roadway was 4 m^2. The rock roadway was in the No. 9 coal seam and about 60 m away from the roof of the No. 15 coal seam. The horizontal distance from the rock roadway to the air return roadway was 50 m corresponding to the one third of the dip length of the working face. The statistics and the analysis of the intermediate-distance gas drainage of 10 working faces indicated that the average gas flux was 17.1 m^3/min ～ 60.6 m^3/min and the gas concentration was 60%～80%. In principle, shaft sinking methods was arailable to extract the intermediate-distance gas. Gas extraction experiment was done in No. 8103 and No. 8104 working face of Yangquan mining area adopting the method of shaft sinking from the ground. We got some experience from it.

(3) Extraction Methods of Long-distance Gas

Most of the long-distance gas comes from the coal seams in sagging zone. Pressure relief gas has the better condition of flowing in the open crannies along the rock and coal seams and less has the condition of flowing in the crannies through the rock and coal seams because most of the crannies in the sagging zone are open crannies along the rock and coal seams and few of the crannies are through the rock and coal seams. Little of the long-distance gas will effuse into the intermediate-distance area. So it has little negative effect on the safety of the initial seam. The pressure relief coal seam in the long-distance area upper the initial coal seam maybe the main exploited coal seam of the mine under the condition of safe exploitation of coal and gas together in high-gas coal seams. It is necessary for us to adopt an effective gas extraction method in order to make the main exploited coal seam have the safe and efficient exploitation condition.

At present, the successful method of long-distance gas extraction lies in the gridding boreholes through coal and rock seams in the roadway below the floor of the pressure relief coal seams in Huainan mining area. This method will be narrated particularly in the section of the engineering practice of Huainan mining area.

The distribution characteristic of the rock stratum crannies was studied by experiments previously, and it found that the shape of the crannies distribution like an "O" circle. The research results were applied to extract the long and intermediate-distance gas successfully by the method of shaft sinking from the ground. The gas extraction effect was very good (to be continued).

Text 2　Integrated Coal Mining and Methane Extraction (Continued)　煤与瓦斯共采(续)

1. Engineering Practice of Huainan Mining Area in China

The experimental zone was in Dongyi and Dong'er mining section. The initial coal seam was B11 coal seam and the district was 2352(1) working face. The strike and dip length of the working face was 1,640 and 190 m separately. The thickness of the coal seam was 1.5 m~2.4 m and the average thickness was 2 m. The slope angle of the coal seam was 6°~13° and the average angle was 9°. The gas content of B11 coal seam was 4 m^3/t~7.5 m^3/t, so it had no coal and gas outburst danger. The thickness of the B11 coal seam was even and the geological structure was simple. The intending yield from the working face was 2,000 t every day when adopting mechanized exploitation. The pressure relief coal seam was C13 coal seam and the district was 2121(3)/ 2322(3) working face. It was about 70 m away from the roof of the B11 coal seam. The strike and dip length of the working face was 1,680 and 160 m separately. The thickness of the coal seam was 5.57 m~6.25 m and the average thickness was 6 m. The slope angle of the coal seam was 6°~13° and the average angle was 9°. The measured gas pressure of the C13 coal seam in this district was 4.4 MPa, the gas content was 13.0 m³/t and the initial permeability coefficient was 0.011 m²/(MPa². d). Many coal and gas outburst and gas explosion accidents happened in the C13 coal seam. The thickness of the C13 coal seam was even and the geological structure was simple. After we adopted the methods to relieve pressure and eliminate the coal and gas outburst danger. The intending yield from the working face was 5,000 t every day when adopting integrated caving mechanized exploitation. Table 10.1 shows the brief condition of coal seams in experimental zone.

Table 10.1 Brief condition of coal seams in experimental zone

Coal seam	Thickness of coal seam/m	Interval of coal seam/m	Comparative interval of coal seam/B	Gas content/(m³/t)	Horizon relation of the pressure relief coal seams
C13-2	0.9	77	38.5	13.0	long-distance pressure relief coal seam
C13	6.0	70	35	13.0	long-distance pressure relief coal seam
C12	0.8	66	33	13.0	long-distance pressure relief coal seam
B11-2	0.4	2	1	5.5	short-distance pressure relief coal seam
B11	2	0	0	5.5	Initial coal seam

(1) Technical Project of Safe High-efficient Exploitation of Coal and Gas

The B11 coal seam with low gas content and no coal and gas outburst danger was exploited firstly. Then the pressure on the C12 and C13 coal seams (in the sagging zone) that were 70 m above the B11 coal seam was relieved and some crannies were formed making use of the exploiting effect. So the active condition of desorption-diffusion-infiltration flow of pressure relief gas was formed. The gas effusion from the long-distance pressure relief coal seam had little effect on the production of the initial coal seam because most of the crannies in the sagging zone were open crannies along the rock and coal seams and few of the crannies were through the rock and coal seams. But because the C13 coal seam was the main exploited coal seam, so we adopted the high efficient method of long-distance gas extraction to reduce the gas content greatly in pressure relief zone, eliminate the outburst danger and achieve the technical demand of integrated caving mechanized exploitation.

(2) Technical Project of Long-distance Gas Extraction

We chose the technical project of gridding boreholes through coal and rock seams in the roadway after a detailed technical and economic comparison below the floor of the pressure relief coal seams to extract the long-distance gas by that the gas extraction roadway was a detailed technical and economic comparison. It was illustrated in Fig. 10.3(a), gas extraction roadway was disposed along the strike direction of the rock seam in the middle of the projection of the haulage roadway and air return airway of the pressure relief coal seam working face.

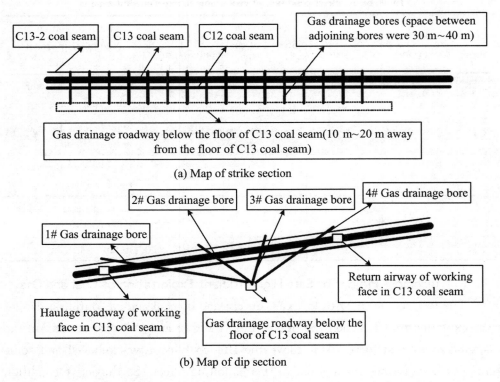

Fig. 10.3　Sketch map of method of long-distance gas extraction by gridding bores through coal and rock seams in the roadway below the floor of C13 coal seam

The gas extraction roadway was in the skewbald claystone and sandstone 10 m~20 m away from the floor of the C13 coal seam. A 5 m long horizontal gas extraction drill site was disposed every 30 m~40 m in the effective pressure relief zone of the initial coal seam. A group of sector bores were drilled in every drill site through coal and rock seams. The diameter of the bore was 0.91 m and the effective gas drainage radius was 15 m~20 m. So the gridding bores through coal and rock seams in the roadway below the floor of the pressure relief coal seam were formed. The released and pressure relief gas in the C13 coal seam accumulated to the borehole along the open crannies under the action of residual gas pressure and extraction negative pressure and then went out to the ground through the gas extraction pipes.

2. Effect of Safe High-efficient Exploitation of Coal and Gas

(1) Gas Extraction Effect

The gas flux began to increase sharply when the working face of the initial exceeded the drill site 40 m away. Fig. 10.4 shows variation of gas flux vs. time in one drill site. The figure reflects the relation between the stress change of coal seam and gas flux clearly. The first 20 days was gas extraction increase period. During this period the stress movement of pressure relieving in coal seam changed sharply and the gas extraction flux increased. From the 20th to the 80th day the gas drainage was active. During this period the stress movement of pressure relieving in coal seam were stable and the permeability was the best. The permeability coefficient increased from 0.011 $m^2/(MPa^2 \cdot d)$ to 32.7 $m^2/(MPa \cdot d)$. It increased nearly 3,000 times. The average gas flux of one bore was more than 1.0 m^3/min and was stable.

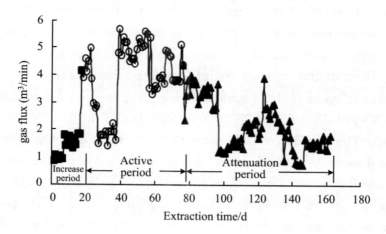

Fig. 10.4 Variation of gas flux vs. time in one drill site

The attenuation period of gas extraction began from the 18th day. During this period the coal seam was compressed tightly gradually, the permeability reduced, the residual gas pressure of coal seam descended and the gas extraction flux descended according to negative exponent. The flux of long-distance gas was 6.5 m^3/min~25.2 m^3/min and the average flux was 16.0 m^3/min. The flux of short-distance gas was 0.9 m^3~12.8 m^3/min and the average flux was 5.0 m^3/min. The high-efficient gas drainage period of the pressure relief coal seam was 2 months under the influence of the exploitation of the initial coal seam. The high-efficient gas extraction strike length was 160 m. 4 drill sites were useful and 16 bores extracted gas simultaneously. The average gas flux was about 1.0 m^3/min every bore. The gas extraction ratio of the coal seam exceeded 60% after the gas was extracted continuously 4 months later. The residual gas pressure of the coal seam dropped to 0.5 MPa. The comparative distortion of the roof and roadway was 26.3‰. The parameter of coal and gas outburst during the process of tunneling in the coal roadway and open-off cut was smaller than the prescriptive critical value.

(2) Effect of the Initial Coal Seam Exploitation

The wind supply flux of the 2352(1) working face in the initial coal seam during the mining period was 1,100 m^3/min~1,400 m^3/min. The gas flux along with the wind was 6 m^3/min~14 m^3/min and the average was 9 m^3/min~10 m^3/min. The gas extraction flux was 0.9 m^3/min~12.8 m^3/min and the average was 5.0 m^3/min. The total gas flux was 8.5 m^3/min~24.5 m^3/min and the average was 15.0 m^3/min. The output of working face was 1,400 t/d~2,000 t/d and the average was 1700 t/d. The gas concentration of the return air was 0.5%~1.0% and the average was 0.8%~0.9%. From the above we can conclude that the technical project can realize the safe exploitation of B11 coal seam.

(3) Effect of the Pressure Relief Coal Seam Exploitation

The tunneling of 2121(3) working face in the pressure relief coal seam (C13 coal seam) was behind that of 2352(1) working face in the initial coal seam (B11 coal seam). The time difference between them was more than 4 to 5 months and the strike distance between them was 350 m~450 m. During the tunneling the wind supply flux was 300 m^3/min~400 m^3/min, the gas concentration of the return air was 0.3%~0.7% and the average was 0.5%. The speed of the tunneling in coal roadway exceeded 200 m every month and there wasn't any dynamical phenomenon during the tunneling. During the coal roadway tunneling of the adjacent no pressure relief working face, the index of the gas outburst exceeded the standard greatly. It was necessary to adopt preventing-outburst measures. The speed of the tunneling in coal roadway was only 40 m~60 m every month. It was necessary to adopt the method of advance extraction to prevent coal and gas outburst and reduce the gas effusion of the working face.

The 2121(3) working face in pressure relief coal seam had no coal and gas outburst danger and the gas extraction ratio exceeded 60% because we adopted the methods of pressure relief and long-distance gas extraction. The yield from the working face increased from 1,700 t/d to 5,100 t/d. The comparative gas effusion reduced from 25 m^3/t to 5 m^3/t. The average gas concentration of the return air reduced from 1.15% to 0.5%. The average yield from the working face can achieve 7,000 t/d according to the existing capability of gas extraction and ventilation of the 2121(3) working face in pressure relief coal seam.

Terms 常见表达

瓦斯 gas, methane, firedamp
矿井瓦斯 mine gas
煤层瓦斯含量 gas content in coalseam
瓦斯容量 gas capacity of coal
瓦斯储量 gas reserves
残存瓦斯 residual gas
煤层瓦斯压力 coalbed gas pressure
吸附等温线 adsorption isothermal
吸附等压线 adsorption isobar
瓦斯涌出 gas emission
矿井瓦斯涌出量 (underground) mine gas emission rate
绝对瓦斯涌比量 absolute gas emission rate
相对瓦斯涌出量 relative gas emission rate
瓦斯矿井等级 classification of gaseous mine
瓦斯等量线 gas emission isovol
高瓦斯矿井 gassy mine
低瓦斯矿井 low gaseous mine
煤(岩)与瓦斯突出 coal(rock) and gas outburst mine

煤(岩)与瓦斯突出矿井 coal (rock) and gas outburst mine
瓦斯爆炸 gas explosion
瓦斯爆炸界限 gas explosion limits
瓦斯层 firedamp layer
瓦斯风化带 gas weathered zone
瓦斯梯度 gas gradient
瓦斯预测图 gas emission forecast map
瓦斯喷出 gas blower, gas blow out
突出强度 intensity of outburst
震动爆破 shock blasting
水力冲孔 hydraulic-flushing in hole
保护层 protective seam
被保护层 protected seam
瓦斯抽放 gas drainage
瓦斯抽放率 gas drainage efficiency
煤层透气性 gas permeability of coal seam
渗透率 infiltration rate
达西 Darcy
达西定律 Darcy's law

Unit 11 Surface Mining

露 天 开 采

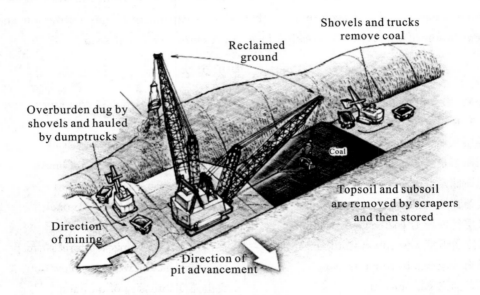

Fig. 11.1 Equipment layout of surface mining

Surface mining is a type of mining in which soil and rock overlying the mineral deposit (the overburden) are removed. It is the opposite of underground mining, in which the overlying rock is left in place, and the mineral removed through shafts or tunnels. In most forms of surface mining, heavy equipment, such as earthmovers, first remove the overburden. Next, huge machines, such as dragline excavators or bucket wheel excavators, extract the mineral.

Text 1 Introduction to Surface Mining 露天开采概论

Surface mining (also commonly called strip mining, though this is actually only one possible form of surface mining), is a type of mining in which soil and rock overlying the mineral deposit (the overburden) are removed. It is the

Unit 11 Surface Mining 露天开采

opposite of underground mining, in which the overlying rock is left in place, and the mineral removed through shafts or tunnels.

Surface mining began in the mid-16th century and is practiced throughout the world, although the majority of surface mining occurs in North America. It gained popularity throughout the 20th century, and is now the predominant form of mining in coal beds such as those in Appalachia and America's midwest.

In most forms of surface mining, heavy equipment, such as earthmovers, is firstly put in place to remove the overburden. Next, huge machines, such as dragline excavators or bucket wheel excavators, will be able to extract the mineral.

1. Ore Reserves Suitable for Surface Mining

Ore reserves suitable for surface mining can be classified initially as:

① Relatively horizontal stratified reserves with a thin or thick covering of overburden; ② Stratified vein-type deposits with an inclination steeper than the natural angle of repose of the material so that waste cannot be tipped inside the pit; ③ Massive deposits, deep and very large laterally such that dumping of the waste within the pit is not possible.

2. Different Types of Surface Mines

Of all the variations of surface mining methods available, the three most common methods only will be described here, namely: strip mining, terrace mining and open-pit mining.

(1) Strip Mining

Strip mining is ideally applied where the surface of the ground and the ore body itself are relatively horizontal and not too deep under the surface, and a wide area is available to be mined in a series of strips.

Walking draglines are for many years the most popular machine for this type of mining due to their flexibility, utility and availability, but more importantly, their low operating costs for waste mining. The dragline is a typical combined cyclic excavator and material carrier since it both excavates material and dumps it without the use of trucks or conveyor belts. The dragline sits above the waste or overburden block, usually 50 m or so wide, on the highwall side and excavates the material in front of itself, to dump it on the low-wall or spoil side of the strip to uncover the coal seam below it.

For maximum productivity, a long strip is required (over 2 km in length) to reduce excessive "dead-heading" time. Longer pits increase the risks of time dependant slope failure in both the highwall and the (waste) lowwall and take up large surface areas that can cause rehabilitation and transport problems. If mixing of coal is important (to meet sales specifications) then long stripping lengths are also problematic in terms of the active mining fronts available for mixing the coal. Where highwall or low-wall stability is problematic, it becomes necessary to monitor the stability of the pit extensively.

Nowadays, several large strip mines operate in areas that were previously mined by underground methods. In such cases it is difficult to anticipate the stability of the overburden and geotechnical surveys are required especially where underground rooms are required to be blasted (by collapsing the pillars between) prior to using a dragline on these areas. Irrespective of the precise geology of the coal seam, the general approach remains similar to conventional strip mining and Fig. 11.2 shows the terminology used of strip mining.

(2) Terrace Mining

When the overburden is too thick (or the floor of the pit is too steeply dipping) to allow waste dumping directly over the pit (as is the case with a dragline and strip mining), it is necessary to use intermediate cyclic or continuous transport (e.g. trucks or conveyors) to transport the overburden to where it can be tipped back into the previously mined void.

It is a multi-benched sideways-moving method, the whole mine is moved over the ore reserve from one end to the other, but not necessarily in a single bench. The number of benches used is usually a function of the excavation depth and type of machinery used (typically between 10 m~15 m benches height and 1 m~32 m benches in the terrace).

Examples of this type of mining can be found in the German lignite mines (where bucket-wheel excavators are used to excavate the overburden — a typical example of a continuous excavating system) and, to a lesser extent, some coal mines in the UK. In these cases, trucks and shovels are used to work 10 benches simultaneously to expose the coal seams underneath. The uppermost layer of overburden is normally mined using hydraulic excavators and trucks, or (when soft material exists), using a bucket wheel excavator, conveyor belt and stacker. These methods are more expensive to use than a dragline, but the dragline is itself not suited to this type of mining due to the limited dump radius of the machine and the much larger width of a terrace mine compared to a strip mine. It is possible however to use a dragline in combination with a "long" terrace as illustrated in Fig. 11.3, but only in the lower or bottom benches where the dig and dump point are within the working radius of the dragline.

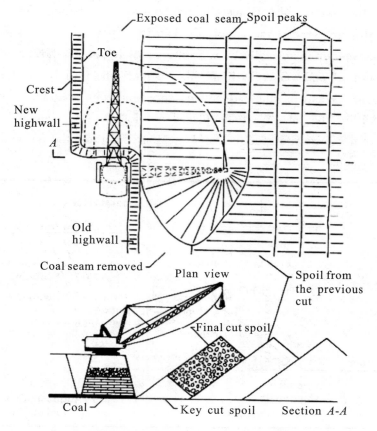

Fig. 11.2　Strip mining terminology

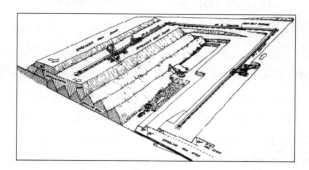

Fig. 11.3　Combined terrace (pre-stripping of soft overburden) and strip mining (stripping deeper hard overburden) methods and associated equipment

(3) Open-pit mining

This is the traditional cone-shaped excavation (although it can be any shape, depending on the size and shape of the ore body) that is used when the ore body is typically pipe-shaped, vein-type, steeply dipping stratified or irregular. Although it is most often associated with metallic ore bodies, it can be used for any deposit that suits the geometry — most typically diamond pipes.

The excavation is normally by rope-shorels or hydraulic shovels with trucks carrying both ore and waste. Drill and blast is most often used, which makes the process cyclic. Waste is dumped outside the mined-out area since no room is available within the pit. Waste is placed as close to the edge of the pit as possible, to minimize transport costs. Fig. 11.4 illustrates the terminology used in the pit design.

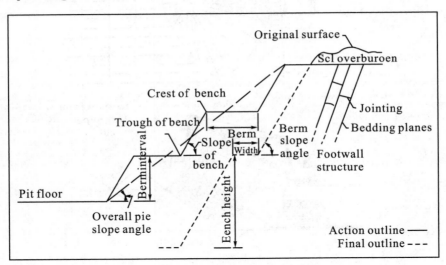

Fig. 11.4 Typical open-pit bench terminology

Benches are normally excavated from 2 m~15 m in height in stacks of 3 to 4, in between which is a crest on which the haul road is placed. When the number of benches in the stack increases, the road gradient increases too. Benches in the stack have a steep face angle whilst the stack and overall slope angles are flatter, thereby helping to prevent slope failures. From an analysis of overall slope geometry, it is clear that as steep a slope as possible should be mined, to reduce the overall stripping ratio. However, this rule is limited by the maximum gradient of the haul road—typically 8%~10% of which requires frequent wider crests, and is to meet the need to have flatter slope angles in places to provide

Unit 11 Surface Mining 露天开采

slope stability. But please note that each pit slope can have a different angle according to the requirements of the design — with or without haul road, geology, etc.

Text 2 Surface Coal Mining Methods in China 中国的露天煤矿开采方法

1. Introduction

China is the largest coal producing country in the world with annual production rate of 324 Mt in the year 2010 of which surface mining production is only about 9% of the total production because the coal reserves suitable for surface mining is not large enough compare with the other major coal producing countries. Surface coal mining methods used in China have also been limited because of the geological, topographical and coal occurrences conditions.

Surface mining of coal was started around 1900 in China and in 1949 (the founding year of the People's Republic of China) the total production of coal was only 32.43 Mt, in the longer period of coal extraction from the beginning of the 20th century to 1980s, surface coal mining production was no more than 4% of the whole country's production rate because the most coal beds are not suitable for surface mining methods to exploit. By the end of the year 2010, the annual coal production rate reached 3.24 billion tons, 9% of which surface mined production was about 9%.

2. Case Study: Antaibao Surface Coal Mine

Antaibao surface coal mine is located in Shuozhou, Shanxi province, started operation in 1985, with designed production rate of 15 Mt / a dry raw coal. The average mineable thickness of the coal seams is 7 m for No. 4 seam, 14 m for the No. 9 seam, and 3 m for the No. 11 seam. The cover over the No. 4 seam varies from 80 m to 120 m thick. The interburden between No. 4 and No. 9 seam is 35 m to 45 m thick, and the parting between the No. 9 and the No. 11 seam is 6 m to 10 m thick. The average seam dip as a result of an anticline trending northeast and plunging southwest is from 2° to 6°.

Fig. 11.5 shows the total reserve area and the mining pits with the direction of mining indicated. Excavation of pit 1 occurs in two phases: development and full production. The development phase is the mining work done prior to the preparation plant beginning full production. The full production phase is the time required to complete Pit 1 and to establish a normal haul back operation mining north in Pit 2.

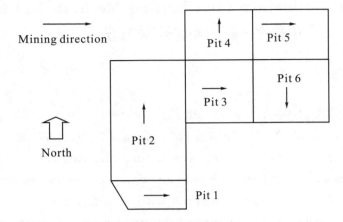

Fig. 11.5 Antaibao surface coal mine mining areas and pits

During the development phase, all the excavated material must be hauled out of the pit to the waste area or the coal stock pile area. The full production phase of Pit 1 requires the waste material above the No. 4 seam to be hauled to the waste areas outside of the pit and the waste bellow the No. 4 seam to be backfilled in the mined out area of Pit 1. The coal is hauled to the raw coal dump at the plant site by the 154-tonne end dump trucks.

The excavation equipment includes the bucket of 25 m^3 in size to be used in overburden removal, large hydraulic shovels, front end loaders in various sizes selected for coal loading and miscellaneous work types.

Fig. 11.6 Antaibao surface coal mine in operation

Unit 11 Surface Mining 露天开采

Terms 常见表达

露天采场 open-pit
山坡露天采场 mountain surface mine
凹陷露天采场 pit mine
露天开采境界 pit limit
（露天采场）边帮 pit-slope
最终边帮 final pit slope
工作帮 working slope
非工作帮 non-working slope
边帮角 slope angle
（露天采场）底面 pit bottom
地表境界线 surface boundary
底部境界线 floor boundary line
封闭圈 closed level
底帮 foot slope
顶帮 top slope
端帮 end slope
剥离 stripping
剥离物 overburden, spoil, waste
剥采比 stripping ratio
平均剥采比 overall stripping ratio
境界剥采比 pit-limit stripping ratio
经济剥采比 economic stripping
剥离高峰 peak of stripping
生产剥采比 operational stripping ratio
剥采比均衡 stripping balance
排土场 dump
外部排土场 external dump
内部排土场 inner dump
台阶 bench
平盘 berm
台阶高度 bench height
台阶坡面 bench slope
坡顶线 bench edge
台阶坡面角 bench angle

工作平盘 working berm
运输平盘 haulage berm
安全平盘 safety berm
清扫平盘 cleaning berm
露天矿开拓 surface mine development
外部沟 external access
内部沟 internal access
双侧沟 double-side access
单侧沟 hillside ditch
缓沟 easy access
陡沟 steep access
开段沟 pioneer cut
坑线 ramp
固定坑线 permanent ramp
移动坑线 temporary ramp
折返坑线 zigzag ramp
回返坑线 rup-around ramp
直进坑线 straight ramp
螺旋坑线 spiral ramp
开采程序 mining sequence
直进坑线 straight ramp
螺旋坑线 spiral ramp
开采程序 mining sequence
分区开采 mining by areas
分期开采 mining by stages
工作线 front
平行推进 parallel advance
扇形推进 fan advance
采场延深 pit deepening
（剥离）倒堆 casting
再倒堆 overcasting
纵向移运 longitudinal removal
横向移运 cross removal
间断开采工艺 discontinuous mining

technology
连续开采工艺 continuous mining technology
半连续开采工艺 semi-continuous mining technology
综合开采工艺 combined mining technology
矿岩准备 preparation of materials
垂直切片 terrace cut slice
水平切片 dropping cut slice
采装 loading
采掘带 cut
采掘区 block
采宽 cut width
工作面 working face
上挖 up digging
下挖 down digging
平装 level loading
上装 upper level loading
下装 lower level loading
满斗系数 bucket factor, dipper factor
挖掘系数 excavation factor
车铲比 truck to shovel ratio, train to shovel ratio
车铲容积比 volume ratio of truck to dipper, volume ratio of train to dipper
组合台阶 bench group
固定线路 permanent haulage line
半固定线路 semi-permanent haulage line
移动线路 shiftable-haulage line
剥离站 waste station
采矿站 ore station
折返站 switchback station
分流站 distribution station
限制区间 limit section
限制坡度 limit grade

入换 spotting, train exchange
入换站 exchange station
排土 dumping
排土场下沉系数 subsidence factor of dump
露天矿生产能力 capacity of surface mine
露天矿采剥能力 mining and stripping capacity of surface mine
水力剥离 hydraulic stripping
逆向冲采 contrary efflux
顺向冲采 longitudinal efflux
水力排土 debris disposal
水力排土场 debris disposal area
水力崩落 hydraulic loosening
滑坡 slope slide
边帮稳定性 slope stability
边帮安全系数 slope safety factor, slope stability factor
边帮破坏 slope failure
塌落 fall
倾倒 toppling
滑(坡)体 sliding mass
滑(动)面 sliding surface
平面型滑坡 plane sliding
圆弧型滑坡 circle sliding
楔体型滑坡 wedge sliding
临界滑面 critical sliding surface
边帮加固 slope reinforcement
边帮监测 slope monitoring
滑坡预报 slide prediction
排土场滑坡 dump slide
排土场泥石流 dump mud-rock flow
挖掘机 excavator
单斗挖掘机 single-bucket shovel
多斗挖掘机 multi-bucket excavator, continuous excavator

Unit 11　Surface Mining　露天开采

正铲(挖掘机) face shovel
反铲(挖掘机) backhoe shovel, ditcher
拉铲(挖掘机) dragline excavator
轮斗挖掘机 (bucket) wheel excavator
链斗挖掘机 chain excavator
露天采矿机 surface miner
螺旋采煤机 auger miner
(悬臂)排土机 spreader, stacker
运输排土桥 conveyor bridge
推土犁 dumping plough
横向排运机组 cross-pit system
铲运机 scraper
矿用卡车 mine truck
架线辅助式矿用卡车 trolley-assisted mine truck
带式输送机移设机 conveyor track shifter

Unit 12　Clean Coal Technology and Environment Protection

洁净煤技术与环境保护

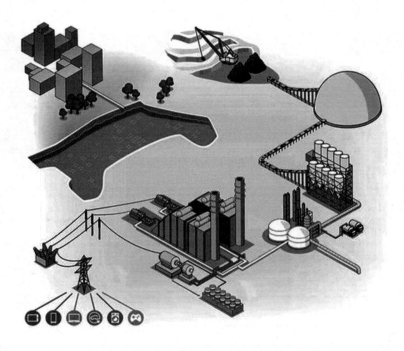

Fig. 12.1　Schematic diagram of clean coal technology

Text 1　Coal Mine and the Environment　矿山与环境

　　Earth is a closed system. Anything that is changed on Earth causes a series of changes in the land, water and air. Coal is the least clean fossil fuel with respect to both local and global environment issues. When coal seams are disturbed, changes will occur in Earth's systems too. Mining coal releases dust and gas into the air. Water is used to wash impurities from the coal. The surface of the land may be changed a lot or a little. The environmental impacts include those of the mining industry and coal transportation — on the landscape, rivers, water tables and other environmental media. This text, however, focuses only some impacts of coal mining to the environment.

1. Air Quality and Greenhouse Gas Concentration

Coal combustion emits particulates, sulphur oxides, nitrogen oxides, mercury and other metals, including some radioactive materials, in a much higher proportion than oil or natural gas and, therefore, causes local and regional pollution problems (the major cause of acid rain and increased ground-level of zone levels), and global climate change. It generates relatively higher emissions of CO_2 than other fossil fuels, as coal's ratio of hydrogen atoms over carbon atoms and power generation efficiency are relatively low compared to other fossil fuels. Coal is also responsible for methane emissions, notably from mining. While oil accounts for 36% of total primary energy supply (TPES), against 23% for coal, both fuels are responsible for 38% each of global energy-related CO_2 emissions. According to recent IEA projections, based on existing energy policies in both the industrialised and developing world, the share of coal in TPES will fall to 22% and coal will be overtaken by natural gas, but its absolute consumption will continue to increase, at least in the next three decades.

Coal is primarily burnt for electricity generation. Steam coal is also used for processing and comforting heat in many industries and in the residential and commercial sectors. Coal is burnt in isolated stoves or industrial boilers for central heating systems. Coking coal is used in the steel industry. Coal plays a small role in transport, either directly in old steam locomotives in various developing countries, or as a source for liquid fuels (mostly in South Africa). It is also a source of gaseous fuels (synthetic gas). Stronger policies favouring energy efficiency improvements and non-carbon emitting energy sources can modify the landscape — but coal will remain an important energy source in the coming decades. Fuel switching in favour of natural gas is occurring world wide but will be limited by resource availability. In the longer run, while oil and gas will become progressively depleted, coal will remain the largest fossil fuel resource available.

Increased use of coal will exacerbate local, regional and global pollution problems unless cleaner and more efficient coal technologies are used. Ultimately, CO_2 capture and storage could be necessary to reduce global CO_2 emissions. This can find support, for example, by a publication from the USA Department of Energy's Energy Information Administration (EIA 2003).

2. Acid Mine Drainage and Water Pollution

Coal mining operations, active and inactive, bring damages to rivers with acid mine drainage (AMD), sediment, and other toxic pollutants. These pollutants are generated when earth and minerals disturbed during coal extraction coming in contact with the atmosphere, rainwater, runoff, and groundwater.

Polluted water is typically treated by collecting contaminated run off or groundwater in earthen sediment ponds at mining sites. These ponds are designed to slow the rate of discharge, provide time for additional water treatment, if needed, and allow for sediment and other pollutants to settle to the bottom. Most mines have multiple ponds and discharge points, or outlets; some large mines even up to more than 50.

In the case of underground mines, ground and surface water often seep into deep mine workings and are then pumped out and discharged to facilitate the retrieval of coal. Discharge flows are typically controlled by pumps, while discharge quality primarily depends on the makeup of the coal seam and the surrounding rock, or overburden. In some instances, underground pumped discharges from non-acid producing seams are relatively clean and can be directly discharged to receiving streams. In most cases, however, underground mine discharges contain pollutants and are directed to sediment ponds, where they are commingled with surface water and are therefore subject to rules weakened by exemptions when it rains.

3. Impact of Mining on Ecology

As a result of mining, significant areas of land are degraded and existing ecosystems are replaced by undesirable wastes. The mineral extraction process drastically alters the physical and biological nature of a mined area. Strip-mining, commonly practiced to recover coal reserves, destroys vegetation, causes extensive soil damage and destruction and alters microbial communities. In the process of removing desired mineral material, the original vegetation is inevitably destroyed and soil is lost or buried by waste.

4. Noise and Vibration

A cumulative effect of the mining activities like, drilling, blasting, crushing and material transportation, produces huge noise and vibrations in the mining area leading which results in hearing loss, underperformance other health related problems and loss of performance.

Text 2   Clean Coal Technologies   洁净煤技术

Environmental pollution and carbon emission have long been being an evitable problem in the process of coal utilization, thus some scholars and institutions proposed the concepts like "coal elimination in China". However, these ideas don't take into account the actual situation of China's economic and social development, and they can't resolve the issue of clean utilization of fossil fuel either. China has so far solved a batch of the main technical issues on coal clean and efficient utilization, including coal-fired power generation with ultra-low pollutant emissions, high-efficiency green conversion using coal as the raw material, low-cost carbon capture, utilization, and storage (CCUS). In the future, coal resources should be burnt and converted intensively, shifting from primary energy to secondary energy supply and making the coal resource play an equal role in being both raw material and fuel.

1. Clean Coal Technologies

Clean coal is a marketing term often used by the coal industry and coal advocates to describe a group of technologies and industry practices that increase coal-derived energy generation efficiency (including coal gasification), significantly reduce coal — power-plant emissions [including CO_2 through carbon capture and sequestration (CCS)], or convert coal to chemical feedstock or transportation fuels to offset oil demand [for example, by coal to liquids (CTL)]. The application of direct-carbon fuel cells is another method to obtain clean energy for coal but for now is largely confined to the laboratory because commercialization is too expensive and power output too low (~ 1 kW) at this stage of development (Fig. 12.2). From an environmental perspective, coal derived energy is only truly clean with CCS. Variants of some modern coal energy technologies have existed for much of the 20th century, but the low price and relative availability of oil has precluded their widespread adoption. Successfully developed clean coal (with CCS) would allow the USA (or any coal-rich nation) to rely safely on an abundant domestic energy resource. However, according to a report from the Massachusetts Institute of Technology, CCS is not yet guaranteed to work on the scale necessary to contain 90% of the emissions from a major power plant (a DOE goal).

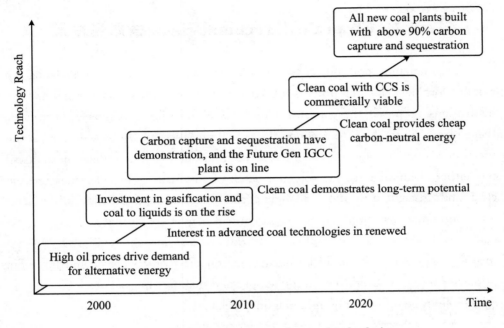

Fig. 12.2 Technology roadmap: clean coal technologies

2. The Enabling Building Blocks

Power plants equipped with CCS can pressurize and pump CO_2 emissions into deep saline reservoirs and depleted oil and gas reservoirs for long-term storage. Pumping CO_2 into oil reservoirs is an established method to enhance oil recovery over the conventional pumping of water and has been under development for that purpose for some time. The Weyburn Enhanced Oil Recovery Project in North Dakota and Canada has used CO_2 from a regional coal-gasification plant to enhance oil extraction since 2000.

Adding CCS remediation to a coal power plant at present consumes about 40% of the power that the plant produces and increases the cost of the energy it produces by 2.7¢/(kW·h), and can not be operated on the scale necessary to collect a majority of the GHG. The largest CCS project in operation today (the Sleipner gas field in the North Sea) sequesters 1 Mt of carbon dioxide per year, which is a small fraction of that generated by a coal-fired power plant, hence CCS development has a long way to go before it can reach the DOE's carbon-capture goal.

Integrated gasification combined cycle (IGCC) provides improved energy recovery from coal versus burning the coal to drive an electricity-generating turbine with pressurized steam. By heating the coal under an oxygen and water atmosphere (no nitrogen), the gasification process generates selected combinations of product, including heat energy, carbon monoxide, hydrogen, methane, and carbon dioxide. The carbon monoxide or methane can serve as a chemical feedstock or burn completely to carbon dioxide. Similarly, an IGCC plant can collect hydrogen as an added fuel product or power an additional gas-driven generator to produce electricity. Remaining solids can find use in a conventional coal-burning furnace as a low-grade fuel. The leftover mineral components are often recovered as useful industrial materials much like flyash is recovered from coal-burning plants for use in concrete (the process is beneficiation).

Elimination of the nitrogen (normally, 80% of air) means that the CO_2 produced by the plant is fairly pure and is prime for sequestration. Besides increasing emission cleanliness and useful recoverable materials, IGCC plants can also generate power about 20% more efficiently than coal-burning plants.

Room exists also for improved efficiency of operation in conventional coal-burning plants. Pulverized-coal — (PC) burning power plants can sustain energy-efficiency improvements through increased temperature of operation. Some new boiler designs also include fluidized-bed operation in which the coal is suspended with a flow of pressurized gas (making it seem somewhat like quicksand). Increased surface contact between coal and oxidizing gases increases furnace temperature, and the fluidized bed allows for noncombustible materials to settle as a slag for potential beneficiation. These attributes increase efficiency and ease elimination of various pollutants from plant emissions (including sulfur).

Coal-to-liquids technology was widely developed by Germany under the fuel embargo leading up to World War II, and apartheid South Africa followed similar embargos lasting much of the second half of the 20th century. Today, South Africa's Sasol (South African Coal and Oil) operates one of the few profitable coal-to-liquids operations in the world, providing fuel and chemical feedstock to

South African industry and for export. Although coal-to-liquids technologies do not reduce greenhouse-gas emissions relative to petroleum (in fact, credible studies show that CTL may increase GHG emissions), they do provide opportunity for coal producers to diversify their product's utility and for coal-rich nations to depend less on petroleum and chemical products that derive from oil, diluting the geopolitical strength of oil-producing nations.

At present, coal capture, utilization and sequestration (CCUS) technology is still at the start-up stage in China. Fig. 12.3 shows the technology principles and roadmap of CCUS. Associated research has been conducting actively by Chinese universities and institutes as well as many other industries related to fossil fuel, such as petroleum exploitation, coal-fired power, and coal-to-liquids. Research and industrial demonstrations have been conducted in many aspects, such as CO_2 capture and utilization, oxygenen riched combustion, high-purity CO_2 geological sequestration, enhanced oil recovery (EOR) and enhanced coal bed methane recovery. Huaneng Group took Beijing thermal power plant as a demonstration project of CO_2 capture. Completed and put into operation in 2008, it successfully captured the CO_2 with a purity of 98% and achieved over 85% recovery rate and 3,000 t/a recovery amount. The CO_2 captured can be refined to the food-grade purity, which can be used in beverage and food industries. Huaneng Group also pushed forward the implementation of a CO_2 capture demonstration project with a capacity of 100,000 t/a at Shidongkou second power plant in Shanghai. Furthermore, In this regard, Huaneng Group's Tianjin IGCC power plant demonstration project is under construction where the CO_2 captured is used for EOR. In terms of EOR, China started on CO_2-EOR/CO_2-EGR in 2005, which is relatively late. China National Petroleum Corporation (CNPC) conducted the filed trail of CO_2-EOR in Daqing oil field in 2008 and Sinopec built a 30,000 t/a CO_2-EOR demonstration project in Shengli oil field. China cooperated with Canada on the first research project about CO_2-EGR applied in deep coal seams, and its current test results show that injecting CO_2 into the coal bed methane well will increase the methane recovery rate. Furthermore, the research data shows 1 Mt of CO_2 can be sequestrated into one square kilometer coal seam, and the recovery rate of coal-bed methane well can be improved by 80%.

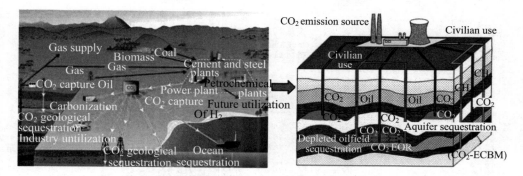

Fig. 12.3 CO_2 capture and geological sequestration

Text 3　Plan of Clean Coal Utilization　洁净煤利用规划

"Technical Revolution in Ecological and Efficient Coal Mining and Utilization & Intelligence and Diverse Coordination of Coal-based Energy System" initiated by the Chinese Academy of Engineering, puts forward three stages (3.0, 4.0 and 5.0) of China's coal industry development strategy. The China Coal Industry 3.0, from now to 2025, strives to fulfill "reduced staff, ultra-low ecological damage, and emission level near to natural gas" through 10 years' technology progress and transitional development. Furthermore, China Coal Industry 4.0 plans to realize "near-unmanned mining and near-zero emission" in 2035. Finally, China Coal Industry 5.0 may reach the objective "no coal above ground, no staff underground, zero emission and zero damage", completing the transformation from traditional energy towards clean energy. In detail, the development of coal mining and utilization over the next ten years is analyzed and three key technologies during China Coal Industry 3.0 (2016~2025), including intelligent coal mining, ecological mining, ultra-low emission and environmental protection, is proposed.

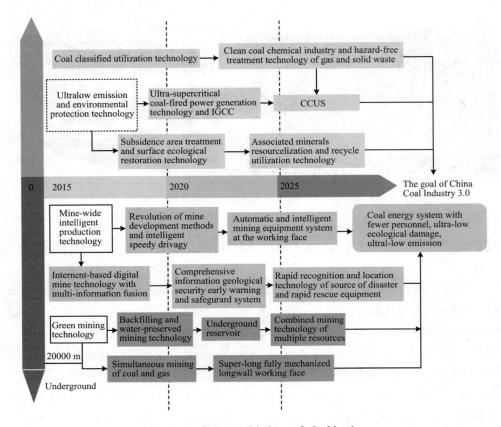

Fig. 12.4　China coal industry 3.0 objectives

Terms　常见表达

选煤 coal preparation, coal cleaning
毛煤 run-of-mine, ROM, r. o. m coal
原煤 raw coal
外来煤 foreign coal
原料煤 feed coal
选煤厂 coal preparation plant, washery
筛选厂 sizing plant
矿井选煤厂 pithead coal preparation plant
群矿选煤厂 groupmine's coal preparation plant
矿区选煤厂 mine coal preparation plant

中心选煤厂 central coal preparation plant
用户选煤厂 user's coal preparation plant
分选作业 separation process
辅助作业 auxiliary process
工艺原则流程图 basic flowsheet
工艺流程图 process flowsheet
设备流程图 equipment flowsheet
粒度 size
入料上限 top size
入料下限 lower size
分选粒级 size range of separation
可选性 washability, preparability

浮沉试验 float-and-sink analysis, float-and-sink test
浮物 floats
沉物 sinks
密度级 densimetric fractions, density fractions
密度组成 densimetric consist, density consist
可选性曲线 washability curves
灰分特性曲线 characteristic ashcurve, elementary ash curve
浮物曲线 cumulative floatcurve, cumulative ash float curve
沉物曲线 cumulative sinkcurve, cumulative ash-sink curve
密度曲线 densimetric curve, yield-densimetric curve, relative density curve
泥化 degradation in water
煤泥(粉)浮沉试验 fine coal float-and-sink test, fine coal float-and-sink analysis
分级 classification, sizing
水力分级 hydraulic classification
水利分析 hydraulic analysis
沉降末速 terminal velocity
等沉粒 equal falling particles
等沉比 equal falling ratio
自由沉降 free falling
干扰沉降 hindered falling
煤矿矿区环境 coal mine environment
煤矿地质环境 geological environment of coal mine
煤矿地下环境 underground mining environment
矿区生态破坏 ecological deteriora mining area

开采损害 mining-induced environmental damage
矿区土地破坏 land deterioration in mining area
矿区水资源破坏 water resources deterioration in mining area
(土地)复垦 land reclamation
矿区(土地)复垦 subsidence trough reclamation
矸石山复垦 waste heap reclamation
生物复垦 biological reclamation
矿区绿化 plantation in mining area
煤矿环境污染与防治 environmental pollution and prevention in coal mine
煤矿环境污染 environmental pollution in coal mine
矿区大气污染 air pollution in mining area
燃煤污染 coal burning pollution
煤烟型大气污染 airpollution due to coal combusion
烟尘 flue dust
消烟除尘 smoke prevention and dust control
烟气脱硫 flue gas desulfurization
矿区水(体)污染 mining area water pollution
选煤废水 coal preparation waste water
矿井水 mine water
高矿化度矿井水 highly-mineralized mine water
酸性矿井水 acid mine water
露天矿坑水 surface mine water
矿井水资源化 reclamation of mine water
煤矿固体废物 coal mine solid waste
矸石山自燃 spontaneous combustion of waste heap

矸石山喷爆 explosion and blower of waste heap
矸石山淋溶水 leaching water from waste heap
矸石处置 waste disposal
煤矿噪声 noise in coal mine
矿区景观破坏 visual impact in mining area
矿井热害 underground thermal pollution
煤矿环境监测 mine environmental monitoring
煤矿环境影响评价 mine environmental impact assessment
矿区环境规划 mine environmental planning
筛分机 screen
棒条筛 bar screen
条缝筛 wedge-wire screen
弧形筛 seive bend
旋流筛 vortex sieve
振动筛 vibrating screen
共振筛 resonance screen
（振动）概率筛 probability screen
琴弦筛 piano-wire screen, power screen
弛张筛 flip-flop screen
等厚筛 vibrating screen with constant bed thickness
齿辊破碎机 toothed roll crusher
滚筒碎选机 rotary breaker
跳汰机 jig, washbox
空气脉动跳汰机 air pulsating jig
筛侧空气室跳汰机 Baum jig
筛下空气室跳汰机 Batac jig, Tacub jig
跳汰式 jigging chamber
空气室 air chamber
重介质分选机 dense-medium separator, heavy medium separator
斜轮（重介质）分选机 inclined lifting wheel separator
立轮（重介质）分选机 vertical lifting wheel separator
重介质旋流器 dense-medium cyclone
磁选机 magnetic separator
斜槽分选机 counterflow steeply inclined separator
螺旋分选机 spiral
摇床 shaking table, concentrating table
浮选机 flotation machine
机械搅拌式浮选机 subaeration flotation machine, agitation froth machine
喷射式浮选机 jet flotation machine
浮选柱 columned pneumatic flotation machine
煤浆准备器 pulp preprocessor
离心（脱水）机 centrifuge
过滤式离心（脱水）机 basket centrifuge
沉降式离心（脱水）机 bowl centrifuge
沉降过滤式离心（脱水）机 screen-bowl centrifuge
过滤机 filter
真空过滤机 vacuum filter
圆盘式真空过滤机 disc-type filter
压滤机 pressure filter
干燥机 dryer, drier
耙式浓缩机 rake thickener
深锥浓缩机 deep cone thickener
（粉煤）成型机 briquetting machine
冲压式成型机 impact briquetting machine
环式成型机 ring-type briquetting machine
对辊成型机 roller briquetting machine
挤压成型机 single-lead-screw extruding briquetting machine

Unit 13　Mine Safety and Occupational Health

矿山安全与职业健康

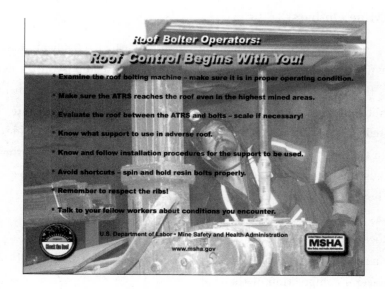

Fig. 13.1　An example of mine safety warning signs

Text 1　Mine Safety　矿山安全

Coal bed gas has been considered a major mine hazard since the first documented coal mine gas explosion in 1810 in the USA. It has considerably affected safety and productivity in underground coal mines throughout the world. Fires are another major safety problem in mines. For example, there were over 7,700 fires during the period between 1947~2006 in underground hard coal mines in Poland alone.

Blasting is an important and hazardous element of mining and many serious injuries and fatalities result from improper judgment or practice during the blasting process. During 1990~1999, approximately 22.3 billion kg of explosives were used by the mining, quarrying, construction, and other industrial sectors in the USA.

Samples of mine atmosphere are usually taken during routine operations to establish a reliable baseline. On the basis of these samples, mine operators can effectively determine changes in mine atmosphere. Furthermore, in the event of a fire-related emergency these samples become very useful in fire fighting operations, rescuing operations, etc.

1. Coal Mine Outburst

It is commonly accepted that by virtue of their chemical and physical characteristics, rock aggregate in the earth's crust are composed of a network of structures that include pores, fractures, and micro cracks filled with liquid and gaseous substances. Coal is one example of these rock aggregate that has accumulated gaseous substances such as methane, carbon dioxide, and/or nitrogen in these structures during the coalification process. Some of the direct or indirect causes identified for outbursts in underground coal mines are as follows:

- The internal energy of the gas contained in the coal bed
- Gas pressure and quantity
- Rock pressure and strength
- Micro seismic activity propagated by reactivation of faults and explosives
- Maceral composition of the coal

Knowledge has accumulated through investigations to determine the origin of coal bed gas and mechanisms of coal mine outbursts has played an instrumental role in enhancing outburst and explosion prediction and prevention methods. Prediction includes activities such as monitoring of gas emissions, monitoring of microseismic acoustic emissions, and monitoring of acoustic signal spectrum characteristics with respect to structures in the rock mass.

Some of the methods that can be used to prevent the occurrence of coalmine outbursts are as follows:

- Utilizing predicting methods

These methods can be used to control and prevent underground coal mine outbursts. For example, by taking gas content and pressure measurements from drill holes on the surface and subsurface, one can determine the threshold conditions for outburst occurrence.

- Modifying mining methods and machinery

This approach is concerned with the modification of the existing methods and machinery to take into consideration rock-mass stresses that could trigger coal bed gas outbursts from an advancing mine face.

- Adopting different methods for gas ventilation or drainage

Two examples of such methods are removing gas through vertical wells in advance of mining, and drilling vertical gob wells into the cave area behind the long wall panel.

Finally, it should be noted that the application of the above methods can be useful to improve mine safety, efficiency of mine operations, and mine economics.

2. Underground Fire

Coal fires are a major safety problem around the globe. Some of the mine-related activities responsible for igniting fires in most hard coal mines are as follows:

- Welding
- Cutting
- Electrical Work
- Explosives

Needless to say, work activities such as these and others have resulted in thousands of underground fires in hard coal mines around the world. For example, in Poland alone there were a total of 7,757 underground fires during the period 1947~2006 in hard coal mines. In spite of many safety-related achievements in combating underground fires in hard coal mines, such accidents still take place now and then.

Out-of-control coal fires around the globe are an environmental catastrophe characterized by factors such as the emission of noxious gases, condensation by products, and particulate matter. Some of the oldest, largest, and environmentally catastrophic out-of-control coal fires in the world are in China, USA, and India. Fires in each of these three countries are discussed below:

- China. China leads the world in coal production, accounting for 1/3 of the global output. A large number of out-of-control coal fires are in the northern part of China, particularly in Xinjiang and Ningxia provinces. Underground coal fires in the Liuhuanggou coalfield in Xinjiang province have been being burning over the past 20 years and the province's capital city, Urumqi, is considered one of the most polluted cities in the world. In the Rujigou coalfield of Ningxia province, underground coal fires are clearly considered responsible for land subsidence as well as for the release of hydrogen sulphide into the atmosphere. All in all, atmospheric pollution throughout China is considered among the highest in the world, and is primarily from coal combustion.

• USA. The USA is the second largest coal producer in the world and has many out-of-control underground coal mine fires, particularly in the state of Pennsylvania. Although Pennsylvania ranks fourth in the USA coal production, after the states of Wyoming, West Virginia, and Kentucky over one third of the USA abandoned mine-associated problems occur in this state and coal fires are considered among the worst such problems. Coal fires in Pennsylvania have been recorded since 1772 and currently there are 140 underground coal mine fires in the state. In particular, the Percy coal mine fire in Youngstown, Pennsylvania has been being burning for more than 30 years.

• India. India is the third largest coal producing country in the world and has many out-of-control coal mine fires, particularly in the state of Bihar. The first coal fire in Jharia coal field at Bhowrah, Bihar broke out in 1916. Currently, there are around 70 fires burning in the Jharia coal field. Most fires in this coal field were due to the spontaneous combustion of coal subsequent to open cast and deep mining, and it is estimated that around 37 Mt of coal have been lost due to the Jharia coal field fires.

3. Explosion

There has scarcely been a major mining industry that has not been traumatized by underground explosions of gases, dusts and mixtures of the two. The potential for disastrous loss of life when such an explosion takes place is very high. Fatality counts have too often been in the hundreds from a single incident. Fatalities and injuries produced by explosions arise from blast effects, burning and, primarily, from the carbon monoxide content of afterdamp - the mixture of gases produced by the explosion. In the dust explosion at Courrieres coal mine in France (1906), 1,099 men lost their lives. One of the most catastrophic explosions on record occurred at Honkieko, Manchuria (1942) when over 1,500 miners died. The carnage that took place in coal mines during the Industrial Revolution was caused primarily by underground explosions and resulted in the start of legislation governing the operation of mines.

An explosion may be defined as a process in which the rates of heat generation, temperature rise and pressure increase become very great due to the rapidity of combustion through the mixture. In a typical methane: air explosion, the temperature rises to some 2,000 ℃, i. e. by a factor of about seven. Even higher temperatures may be reached if the explosion is contained within a sealed volume. The speed of the process is so great that it is essentially adiabatic. The result is that the pressure in the immediate vicinity increases to a peak value very rapidly and is relieved by expansion of the air. This produces a shock wave that propagates in all available directions.

In a mine opening the expansion is constrained by the airway surfaces giving rise to high velocities of propagation. Initially, the flame front travels more slowly than the shock wave and the explosion is known as a deflagration. However, if the unburned zone ahead of the flame front becomes more conducive to combustion by approaching closer to stochastic condition (approximately 9.6% methane) then the flame front will accelerate and the peak pressure at the shock wave rises. Although the shock wave velocity also increases, the distance between the two narrows. This may endure until the flame front catches up with the shock wave (Fig. 13.2). The explosion is then described as a detonation. The speed of the explosion and the peak pressure at the shock wave then escalate significantly. Conversely, if the unburned zone between the flame front and the shock wave moves away from stochastic conditions (e. g. by lack of fuel or the presence of stone dust) then the explosion will weaken.

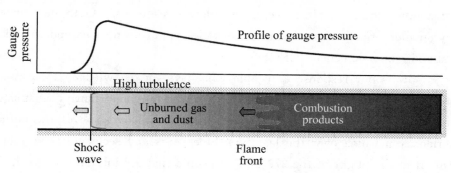

Fig. 13.2 **In an explosive deflagration the flame front is preceded by a shock wave**

The mixing process is further exacerbated by the highly turbulent conditions that exist between the flame front and the shock wave. Such turbulence, coupled with the high speed of the unburned mixture can result in scouring of the sides of the airway, especially if the coal is friable. The impact of flying debris can also result in significant erosion of the airway as well as dispersion of dust into the air.

If there is any obstruction that prevents the air being projected forward in front of the shock wave the peak pressure becomes greatly escalated. This occurs when the shock wave encounters a facing seal, stopping or any other ventilation control. Similarly, high pressures will be generated when an increasingly powerful explosion enters a blind heading. In such circumstances victims are typically found to have suffered from blunt trauma in addition to severe burning. The latter is thought to be caused by swirling flames that exist until extinguished by lack of oxygen.

Text 2　Introduction to Occupational Health and Safety 职业健康与安全概述

The International Labor Organization (ILO) estimates that more than two million people die each year from work related diseases or injuries and more than 270 million people are involved in occupational accidents.

Many of these accidents occur in third world countries where fatalities, occupational diseases and injuries are inherent within the cultural norms. OHS (Occupational Health & Safety) information and the concept of injury prevention and management are often incomprehensible for many small industries that are vital to sustaining social and economic harmony in these countries. While larger organizations embrace OHS concepts and allocate resources to OHS management, many smaller organizations in these countries simply can't yet understand the social and economic benefits of OHS.

So how is our response as people who have the knowledge? How can we share it with those who can't afford to pay for it? More importantly, what can we do to take a world view of OHS? This paper draws on some of the author's experiences in China over the last two-three years and seeks to draw people's attention to the plight of the 270 million people that are injured each year. The author will propose some opportunities for response to these issues from Australian safety practitioners.

1. Introduction — the Australian Context

In the last 10 years Australia has witnessed some significant statistical improvements in occupational health & safety performance. In the Queensland mining industry alone we have seen a 26% reduction in Lost Time Injury Frequency Rates (Palaszczuk, 2005) while increasing the output of the mining industry from an estimated $35 billion to $67 billion or 35% of Australia's GDP. Mining industry fatality rates, while still socially unacceptable, have also fallen significantly to 4% for the 2003~2004 financial year. In a sense, the business community of Australia, Union movements and safety professionals can feel a sense of achievement in these successes.

Statistically, we can appreciate the achievements however, behind each of these incidents, the emotional suffering and social upheaval associated with each is often beyond our personal comprehension. In 2003~2004 year Australia witnessed 144 direct work related fatalities (NOHSC, 2004) and it is probable that work related diseases killed many more. Some authors suggest these figures could be as high as 2,900 per annum. Each one of these fatalities will affect a variety of family and community members in many ways. We are also witness daily to many disabling injuries, which resulted in the 2000 financial year there were 92,900 claims for compensation, which entailed at least 10 days off work for non-fatal injuries or illnesses (ABS, 2000). While the numbers appear somewhat large, they are but a drop in the ocean compared with some of the world's developing economies.

The purpose of this text is 2 fold. Firstly, I will seek to highlight the significant incidence of work related fatalities and disabling injuries in emerging economies that are on Australia's doorstep. Secondly, I will propose ways in which Australian occupational health and safety practitioners might become involved in sharing Australia's experiences in managing occupational health and safety with our neighbors and subsequently decreasing work related fatality rates among our regional neighbors.

2. Workplace Safety has International Impacts

The International Labor Organization (ILO) estimates that more than 2 million people die each year from work related diseases or injuries and more than 270 million people are involved in occupational accidents. That translates down to 740,000 per day, about 514 per minute or around 9 every second. Over the next 25 minutes of this presentation, 96 people world wide will lose their lives in work related fatalities.

Shocking as these statistics are, the news becomes worse. It is projected by the ILO that children account for more than 12 million occupational accidents with greater than 12,000 children killed in work related accidents.

Total mortality rates in developing economies such as China are at least twice as high as those in developed economies like Australia. Data from particular sectors, like the Chinese mining industry indicates that mortality rates are higher still. In China alone the official reported mining death toll is some 6,000 personnel per year (SAWS, 2005), however, some trade union and human rights activists indicate that this figure is somewhat under-reported and may be as high as 12,000~15,000 per year. The Chinese mining industry has reportedly more than 600,000 current cases of coal workers pneumoconiosis and that figure is increasing at a rate of around 60,000 per annum. Reported pneumoconiosis prevalence rates among miners prior to China's recent mining boom, have been estimated by some authors at 6%.

Official statistics indicate that Chinese coal production increased by 9.7% to 940 Mt in the first 6 months of this year. By contrast, the total Queensland Coal mining industry output is expected to be in excess of 170 Mt this year financial year. There are no doubt with such a vibrant economy, there is difficulty in successfully regulating and monitoring the effectiveness of safety regulation. The State Administration of Work Safety (SAWS) and the China National Coal Association are undertaking to reform the industry through rationalization and implementing effective legislative regimes across the industry. However, there are significant unseen barriers to this. These barriers include the reliance of many small villages on local community coal mines for economic and social stability. So to simply rationalize the industry and close down mining operations where small villages rely, and have relied for many years, may result in significant economic hardships for tens of thousands of people. Indeed the reliance of the local coal mine to provide coal for cooking and heating is of paramount to survival in many western areas of China where winter temperatures are well below zero and electricity scarce. With more than 10,000 small mines in China regulation and enforcement is not a simple issue.

Indeed for many hundred of thousands of workers in China, their economic needs are so desperate that they need to work despite the hazardous conditions that may be encountered. This is something that the government is almost powerless to control. A significant number of Chinese coal miners are immigrant workers (i.e. they work away from their home regions, depending on the region and company). Components of the salary are often sent home to family members in their home village. It is reported that in many privately owned mines, workers who refuse to work because of safety related conditions are sacked because there are those that would welcome the opportunity and take their position, albeit hazardous to their life.

3. What Can We Do with this Problem?

When faced with challenges of this magnitude, it may appear somewhat difficult to provide some tangible solutions. It is my belief that there are both individual and corporate responses to take on the challenges of improving safety in developing economies. Individually, we can participate through providing some of our own time to charitable organizations who support development of OHS competence among governments of developing economies and by lobbying Australian aid agencies to make OHS part of their funding considerations.

Corporately we can have some significant impact through influencing social responsibility aspects of our purchasing programs. If we work for a global giant we should consider participating in global programs that ensure OHS equality and practices across all sites not just those in the western world. Ultimately, we should work towards tangible solutions that can directly impact safety. An example of a recent project conducted in China by Szudy, O'Rourke & Brown (2003) was the development of an action based health and safety training project in Southern China. The project focused on developing the capacity of managers and factory workers in respect of OHS in international footwear manufacturers. This was facilitated through training, establishment of workplace health & safety committees, increasing dialogue between buyers (multi nationals) in respect of OHS obligations towards factory workers and the establishment of democratic processes for election of safety representatives. Projects such as these are part of the future solution to improving safety globally.

Our solutions must be effective and provide workers with new knowledge and skills that benefit their well being at work and away from work. And players such as governments, business and enterprise leaders, communities and workers must be able to use that knowledge and those skills to change the way people are put at risk, at work, on roads, in communities and in their homes.

There is no doubt that for individuals, difficulty is going to be meet regarding providing some direct solutions, however, collectively there are benefits from partnerships. Organizations like the ILO, the World Health Organization and the Asia Pacific Occupational Health & Safety Organization provide a range of inputs in minimizing global safety risks.

OSHAID(Occupational Safety & Health Aid) International is a nonprofit organization committed to taking basic occupational health & safety management practices to developing economies. Through the commitment of sponsorships and voluntary participation from individual members, OSHAID International will aim to undertake a range of projects to improve occupational safety in developing economies that have limited capacity to investigate or implement fundamental work safety practices.

Terms　常见表达

外因火灾 exogenous fire
内因火灾 spontaneous fire
煤的燃倾向性 coal spontaneous combustion tendency
自然发火煤层 coal seam proneto spontaneous combustion
自然发火期 spontaneous combustion period
火灾气体 fire gases
火区 sealed fire area
火风压 fire-heating air pressure
均压防灭火 pressure balance for air control
防火门 fireproof door
防火墙 fire stopping
阻化剂 retarder
阻燃 flame-retardation
阻燃剂 flame-retardant agent, fire resistant agent
灌浆 grouting
洒浆 mortar spraying
惰性气体防灭火 inert gas for fire extinguishing

矿山救护队 mine rescue team
呼吸器 respirator
苏生器 resuscitator
自救器 self-rescuer
避难硐室 refuge chamber
自然发火预测 prediction of spontaneous combustion
淹井 mine flooding
矿井突水 water bursting in mine
突水系数 water bursting coefficient
矿井防治水 mine water management
防水闸门 water door
防水墙 bulkhead
矿井堵水 sealing off mine water
注浆堵水 grouting for water-blocking, grout off
煤矿(集中)监测 mine monitoring
矿井环境监测(underground)mine environmental monitoring
矿井火区环境监测(underground) mine fire-zone environmental monitoring
设备健康监测 health monitoring of equipment

Unit 13 Mine Safety and Occupational Health 矿山安全与职业健康

煤矿(集中)监控 mine supervision
煤矿生产(过程)监控 mine productive supervision
煤矿轨道运输监控 mine track haulage supervision
煤矿带式输送机运输监控 minebelt conveyor supervision
采煤工作面监控 coal face supervision
地面数据处理中心 central station
(监控)主站 master station
(监控)分站 outstation
传输接口 transmission interface
星状(监控)传输网 star transmission network
树状(监控)传输网 tree transmission network
环状(监控)传输网 loop transmission network
敏感器 sensor
传感器 transducer
变送器 transmitter
检测仪表 measuring instrument
甲烷传感器 methane transducer
烟雾传感器 smoke transducer
一氧化碳传感器 carbon monoxide transducer
负压传感器 negative pressure transducer
风速传感器 air velocity transducer
煤位计 coal level meter
采煤机位置传感器 shearer position transducer
机车位置传感器 locomotive position transducer
煤岩界面传感器 coal-rock interface transducer
设备开停传感器 equipment on/off transducer
胶带打滑传感器 belt track transducer
胶带撕裂传感器 belt rip transducer
胶带跑偏传感器 belt disalignment transducer
输送机堆煤传感器 conveyor coal blocking transducer
矿用胶带秤 belt conveyor scale for mine
胶带载人保护器 protecting device for belt riding
提升信号装置 hoisting signaling, winding signaling
斜井人车信号装置 inclined shaft manrider signaling
矿区以外通信 communication beyond mining area
矿区通信 mining area communication
矿井通信 (underground) mine communication
井下通信 underground communication
矿井调度通信 (underground) mine dispatching communication
矿井调度通信主系统 main system of (underground) mine dispatching communication
矿井调度通信子系统 subsystem of (underground) mine dispatching communication
矿井局部通信系统 (underground) mine local communication system
调度特权 dispatcher's priority
紧呼 emergency call
选呼 selective call
组呼 group call
全呼 all call
扩音传呼 loudspeaker paging

通播 emergency through broadcasting
工作面通信 face communication
井筒通信 shaft communication
架线电机车载波通信 carrier communication for trolley locomotive
输送机线通信 communication for conveyor line
移动通信 mobile communication
矿山救护通信 mine rescue communication
同线电话通信 party line telephone communication
井下无线通信 underground radio communication
透地通信 through-the-earth communication
感应通信 inductive communication
长感应环线 long inductive communication
漏泄通信 leaky feeder communication
漏泄馈线 leaky feeder
漏泄(同轴)电缆 leaky coaxial cable
漏泄电缆的传输损耗 transmission loss of leaky coaxial cable
漏泄电缆的耦合损耗 coupling loss of leaky coaxial cable
矿用通信电缆 communication cable for mine
矿用光缆 optical fiber cable for mine
低导层固有传播 natural propagation in low conductance layer
巷道固有传播 natural propagation in an empty roadway
巷道的截止频率 roadway cut-off frequency
单线模(式) monofilar mode
双线模(式) bifilar mode
模(式)转换器 mode converter
(调度通信)汇接装置 interconnecting device (for dispatching communication system)
矿用电话机 mine telephone set
声力电话机 sound powered telephone set

Unit 14　Intelligent Precise Coal Mining

煤炭智能精准开采

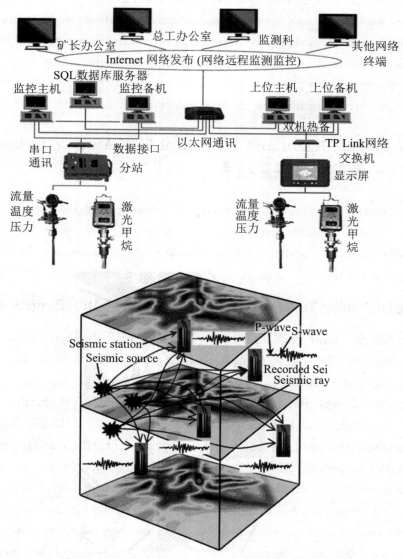

Fig. 14.1　Technical system of intelligent precise coal mining

A scientific concept of intelligent precise coal mining is proposed to meet the challenges and opportunities facing coal mining. By means of technologies including intellisense, intelligent control, the Internet of Things, cloud computing and big data, precise coal mining is proposed as a new future mining mode integrating intelligent mining technique with few workers (unmanned), and disaster prevention and control. This mode is based on transparent spaces and geophysics, as well as multi-field coupling to achieve spatial-temporal accuracy and efficiency. It is able to comprehensively consider factors relating mining under different geological conditions, including mining influences, relevant factors inducing disasters, and ecological destruction caused by exploitation: it is able to enhance the automatisation, intelligentization, and informatization levels of safe coal production which promotes the transformation of the coal industry from a labor-intensive, to a technology-intensive sector.

Text 1 Intelligent and Ecological Coal Mining Technology
  智能生态采矿技术

As the Internet plus era and intelligence development are intensified in the new century, this research investigates how coal mining copes with the advent of a new round of science and technology innovation by summarizing the history of coal mining and the technological development trends therein.

1. Digital Mining Technology of Internet-based Multi-information

At present, most large-scale coal mines have built the digital mine models, which digitizes the early information acquired from the drilling, seismic prospecting, radar scanning and so on, forming the initial coal mine geological data model. Later, the information, such as mining development, roadway layout, ventilation, and safety monitoring and control, is fused. Foundation platforms are structured for the overall production scheduling, safety assurance, transportation, personnel monitoring, shown in Fig. 14.2. However, most of the platforms are only a simple collection and unified managed at present, lacking practical information integration. Therefore, decisions from each level are primarily made by personnel. Meanwhile, because the production control system has no association with the foundation platform, the geological data model can't be dynamically corrected according to production status, and the geological information can't get involved in decision-making either. In the future, for one thing, information acquisition and integration technology should be strengthened by making full use of the existing technology, such as internet of things, automation, intelligence, big data, and cloud computing. For another thing, for the purpose of providing necessary information for intelligent mining and safety assurance, breakthroughs need to be made in the construction and dynamic correction of high accuracy three-dimensional dynamic geological model to build the high accuracy four-dimensional GIS platform.

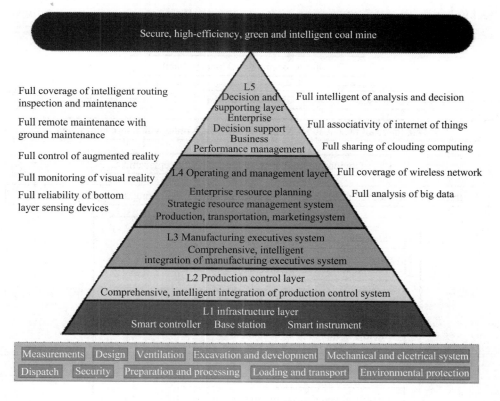

Fig. 14.2 Internet-based multi-information integration digital mining technology

2. Technological Change in Mining Development Method and Intelligent Rapid Excavation

At present, mining development is developing towards large scale and centralization. It is being promoted that only one longwall face for each mine with less chain pillar. Because the gate road development rates generally lags behind those needed to meet longwall capabilities, the conflict between longwall face advance and gate road development has been a prevalent problem, and how to resolve such an issue has been a significant research direction. For the purpose of solving the problem, non-chain-pillar mining method, like "N00 construction method" and gobside entry retaining, has been studied and developed. Furthermore, improving the intelligence and collaboration ability of gate road development system are also an important research area, such as developing integrated equipment that combines driving, bolting, support, transportation, and prospection. Therefore, breakthroughs should be made in the following key technologies: intelligent cutting, fast support, fast transportation, intelligent ventilation, and safety assurance, shown in Fig. 14.3.

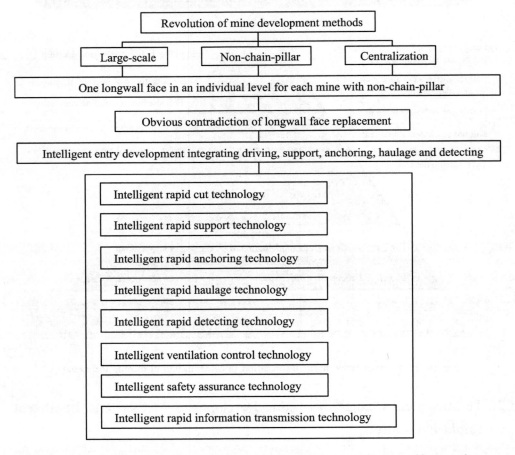

Fig. 14.3 Technological changes in mining development and intelligent rapid excavation

3. Automatic and Intelligent Mining Equipment System at the Working Face

At present, the development trend of fully-mechanized coal mining is automatic and intelligent mining. Automatic technology has been applied to mining engineering around the world since the 1990s, such as tele-mining/remote control mining technology in Canada and automatic mining "Grountecknik 2000" strategy in Sweden. The electro-hydraulic control system of hydraulic roof supports, the automatic control system of the shearer, and other control systems for other equipment have been gradually studied in China since 2000. These systems significantly improve the efficiency and safety of longwall face by the collaborative operation of the shearer, the armored face conveyor (AFC), hydraulic roof supports, visualization of longwall face, and the communication and remote control of longwall face. With the increasing complexity of coal mining, the traditional mechanized and automatic mining technology cannot meet the requirements of further improvement of the mining efficiency and safety level. Therefore, intelligent mining becomes inevitable. LASC straightness monitoring and control system in Australia, as well as new mechanization and automation of longwall and drivage equipment (NEMAEQ) project in Europe, have made significant achievements in the intellectualization of coal mining equipment. Some universities and institutes in China are conducting researches in many fields, such as underground inertial navigation and precise location, straightness intelligent control, intelligent adjustment of cutting height, the achieved results are in the process of perfection and promotion. As shown in Fig. 14.4, the intelligent control technology for one single machine has been resolved, but the control of complete sets of equipment are still centralized control mode based on Industrial Ethernet, which is hard to satisfy the control requirement of rapid increasing fully-mechanized coal mining equipment group. The new control method and architecture innovation on the system level are also gradually developed, including of self-organized cooperative control method of hydraulic roof supports group, dynamic decision-making based on the feedback of sensory perceptual system, longwall face control system and application platform based on distributed decision-making. Once these technologies are successfully applied, the equipment could make correct response to the changes of surrounding rock, significantly improving the level of intelligence of equipment.

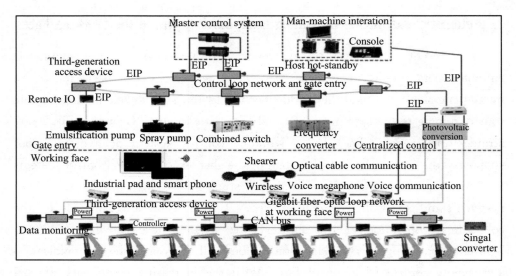

Fig. 14.4 Intelligent equipment and control system of a longwall face

Text 2   Future of China Coal Industry   中国煤炭工业的未来

Through the summary and analysis of aforementioned critical technologies, the great anticipations for the China Coal Industry could be concluded as follows:

• Digital mining, intellectual mining and safety monitoring technology will have an accelerated development, and address technical problems, such as safety situation, mining area environment, working conditions, and labor intensity;

• Continuously improved efficiency of comprehensive utilization of resources, consistently developed ecological mining technology and equipment, and substantially reduced the environmental damage;

• Making an obvious progression in the energy transition from traditional energy to clean energy and making breakthroughs in clean burning coal, high-efficiency conversion, and utilization.

1. "Transparent Earth" — 3D Visualization of Geological Information

Accurate and comprehensive 3D visualization of the geological model is the foundation for exact coal mine planning and development. The China Coal Industry 4.0 and 5.0 need to make breakthroughs in high-precision geological modeling to accurately describe a broader scope of Earth's tectonics and reveal the fundamental laws and dynamic changing characteristics of deep strata deposition.

These breakthroughs can provide essential data support for the exploitation of underground mineral resources and deep ground scientific research.

2. Unmanned Mining

The ultimate objective of automatic and intellectual mining development is the unmanned mining of the entire mining processes. On the basis of artificial intelligence and distributed control, the existing mining and extraction patterns should be completely changed to create a more reliable and intelligent equipment system.

This system should possess the functions of self-decision making, self-control, self-adaption, and self-maintaining and should integrate deep resource mining, processing, back-filling, gasification, power generation, and other subsystems, which will actually realize "No personnel underground and No coal above ground" unmanned mining.

3. Control Technology of Underground Integrated Conversion

In the stages of China Coal Industry 4.0 and 5.0, the clean coal utilization will, hopefully, be able to be completed underground (integrated coal-to-gas, coal-to-liquids, coal-to-power). In other words, the useful energy needed will be transported to the surface while the waste can be dealt with underground. Exact reaction processes and facility parameters are indispensable for the coal conversion. However, the existing technologies can't satisfy the high-efficiency and large-scale conversion processes underground. Therefore, it is inevitable to supersede the current technology and dramatically improve the energy intensity to realize high efficiency and intensive conversion in the limited underground space. Fig. 14.5 is a conceptual model of control technology of precise coal conversion and power transmission.

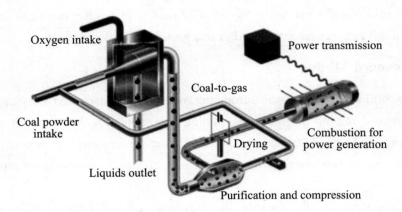

Fig. 14.5 Schematic of precise control technology and process of underground integrated conversion

Unit 15 Comprehensive Utilization of Mining and Mineral Processing Wastes

煤矿固体废弃物综合利用

Mining and mineral-processing wastes generated after mining and ore processing at or near mine sites have no current economic use. A number of environmental problems are associated with the disposal of this waste, including contamination of streams and lakes and pronounced landscape transformation, e. g. , stock-piled waste rock and tailings, subsidence basins, open pits, and removal of overburden rock and topsoil. Despite several efforts to reduce the amount of waste produced, solid mineral wastes remain one of the world's largest waste streams.

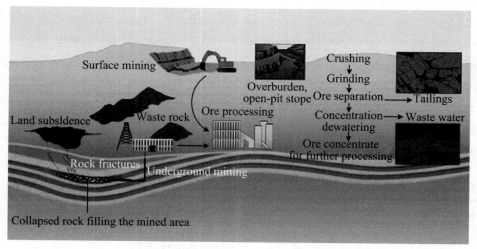

Fig. 15. 1 Environmental problems of coal mining and mineral processing wastes

Text 1 Utilization Techniques of Mining and Mineral Processing Wastes 煤矿固体废弃物利用技术

In order to meet the ever-increasing demands of the modern society, the mineral production is continuously increasing along with the scale of mining operations. However, mineral production is often not in consonance with conservation of environment and forests, which has several adverse impacts including air, water and soil pollution, socio-economic problems and effect on wildlife population and their behaviour. The technological developments in the field of underground mining, viz. mass production equipments, roof support, communication and automation is helping the decision makers to consider underground mining practice for sustainable mining while meeting environmental concerns.

The reuse of mining and mineral-processing wastes may minimize the environmental impacts related to disposal; however, some reuse and recycling measures may actually cause new and serious environmental problems. The overall environmental costs can be determined by various approaches such as ecological risk assessment, lifecycle assessment, sustainability operations assessment, and ecological footprint estimates. Economic cost-benefit analysis, however, is the ultimate driver in terms of the feasibility of a specific reuse technology. If the costs of final target material extraction or mine waste reuse method are economically prohibitive, then even the most eco-friendly process methods will be difficult to implement without regulation or government subsidies.

The question that directs the choice of principles is "what characteristics must a given waste disposal solution exhibit, in the context of local conditions, for mining and mineral processing to proceed, without significantly disturbing the ecosystems, communities and economies overlying and surrounding ore deposits and processing facilities?" The principles are applicable to the various techniques currently applied to manage wastes and the different types of solid wastes produced by mining and minerals processing, including overburden, tailings, waste rock, heap leach piles and in-situ leached rock. The principles provide a set of ideals. In practise, the principles can guide waste disposal decisions through the consideration of what kind of risk and magnitude, in any given local context, may be generated concerning the application of a particular management solution poses to their application.

1. Conventional Tailings Dams

Conventional tailings dams are the most common form of mineral processing waste management utilised by the minerals industry. Tailings dams isolate the waste material from the surrounding ecosystems through storage and containment. Tailings are most commonly piped into dams as a wet slurry with the dam wall progressively built as the volume of waste material increases. Conventional tailings dams are best suited to semi-arid and arid environments, where precipitation does not exceed evaporation. In wetter environments, where sub-aqueous deposition is often practiced as a means of reducing oxidation of potentially acid-generating wastes, high precipitation may lead to excess water that requires discharge and treatment and increase the risks relating to the physical stability of the dam and its containment/bunding structures.

2. Paste and Thickened Tailings

Increasingly, paste and thickened tailings techniques are being used more widely, due to a larger range of thickener technologies, reduced costs, increased familiarity and access to expert knowledge, and increased water scarcity at some localities. Paste and thickened tailings refers to a continuum of tailings with high solid concentrations and higher yield stress, due to the greater level of fluid removal from tailings before disposal. Conventional tailings typically range 30% ~50% solids, thickened tailings 55%~75% and paste over 75% (solid concentrations vary with particle size and shape, clay content, mineralogy, electrostatic forces and flocculant dosing). Paste and thickened tailings can require additional capital expenditure but, over the long-term, thickening techniques may result in decreased management, lower dam construction and rehabilitation costs and significantly lower water use. Thickened tailings require transport by centrifugal pumps, while paste tailings utilise positive displacement pumps and high pressure piping.

3. Direct Disposal

The direct disposal of mining and mineral processing wastes into rivers, oceans and lakes presents significant technical, social and environmental challenges to sustainable development. Riverine tailings disposal (RTD) is the direct discharge of mine process tailings into rivers. Overburden is also sometimes co-disposed during RTD. Riverine tailings disposal is conspicuous and visible and has come to characterise the industry in the public consciousness. The technique is often considered in circumstances, where rugged topography, high rainfall, seismic activity, high groundwater levels, the lack of "cross-valley" locations and the absence of suitable embankment material preclude the impoundment of tailings. The method attracts low up front costs, although is clearly a method of mine waste disposal that has been responsible for cases of serious pollution.

4. Heap Leaching

Heap leaching is a common method for the extraction of metals from ores for some commodities, such as gold and copper. Heap leaching consists of the excavation and crushing of ore, placement on an impermeable membrane and irrigation with a reagent to promote decomposition of ore minerals and mobilisation and capture of the desired metal. The residual piles are a form of mineral processing waste. They may be rehabilitated, but are sometimes left unrehabilitated in the landscape. Heap leach piles face many of the same physical and chemical stability challenges as conventional tailings dams. There are also stability challenges unique to the technique, which include increased erosion, due to the physical form of the material and the often unconsolidated nature of the waste, the long-term integrity of membranes and the contribution of membranes to an increased risk of physical instability of heap leach piles during earthquakes. When heap leach piles are left uncontained and unrehabilitated, there is an increased risk of chemical and physical erosions of the waste. The containment of the irrigation reagents during processing can also present a challenge, particularly for spray-type systems which can be blown onto surrounding ecosystems or communities.

5. In-situ Leaching

In-situ leaching (ISL) is a relatively uncommon mining process technique, whereby reagents are pumped into in-situ ore bodies with the aim to dissolve desired ore minerals into solution for extraction and further processing. ISL is most commonly used for certain types of uranium deposits, but can also be adapted for particular copper or gold deposits depending on the local conditions and issues.

In-situ leached rock can also be considered as a form of mineral processing waste. The leached rock has a high physical stability, as it remains relatively consolidated in its original position (depending on the extent of artificial permeability enhancement). The containment of potential contaminants mobilised as a result of leaching into groundwater, however, can be a challenge for in-situ leaching facilities and a long-term source of contamination. Remediation of contaminated groundwater has proved difficult (Mudd, 2001). Techniques to isolate such wastes may provide opportunities for containment. The return of anoxic conditions post-leaching can reduce the further decomposition of minerals.

6. Wast Rock Dumps and Backfilling

Overburden and waste rock are typically stored in waste dumps surrounding mining operations. Open-cut operations, in particular, generate large volumes of excavated rock that ore deposits overly [overburden; Mudd (2010)]. Some of this rock is non-mineralised (such as sandstone or limestone), while other rocks can contain sulphidic gangue minerals, as well as ore minerals at grades not high enough to be considered ore, and hence not worthy of processing.

In locations of low rainfall, the risk of erosion and the potential for geochemical changes which can cause release of toxic elements may be low and not necessitate additional containment of the waste material. In such circumstances, waste material may remain inert and stable. Wind erosion, however can be a significant mobiliser of contaminants, at some arid localities [e. g. Lottermoser and Ashley (2006)]. In areas where high rainfall and/or high intensity rainfall events can occur, erosion may lead to the interaction of wastes with the environment. Waste rock containing metal sulphides presents particular challenges for responsible waste management in such environments. Following excavation and exposure of sulphide ores to the vastly different surficial environmental conditions, the minerals can oxidise and produce undesirable decomposition products. This process is the inevitable result of the re-equilibration of minerals formed at high temperatures to weathering conditions on the surface. During this process, potential contaminants can be released and mobilised, particularly in the acidic environment often created during the dissolution of the sulphides. This process is commonly known as acid rock drainage (ARD), acid mine drainage or acid and metalliferous drainage (AMD)

(Akcil and Koldas, 2006, Department of Industry, Tourism and Resources, 2007). It should be noted that acidic conditions are not necessarily the only conditions under which metals in dissolution are mobile. The neutralisation of the acid by dilution or reaction with, for example, carbonates (limestone) may not halt the mobilisation of some contaminants. Apart from those occurring in the driest of environments, waste dumps usually require techniques, such as capping and rehabilitation with vegetation, to minimise water infiltration. In those circumstances where deep drainage through mineralised wastes still occurs, ongoing treatment of seepage by various active and/or passive methods may be required.

Overburden, waste rock and dried tailings may also be placed into mining voids to reduce the surface footprint and demand for surface dumps, a technique referred to as dry backfilling (Dixon-Hardy and Engels, 2007). Dry backfill can exhibit high physical and chemical stabilities and containment in a dramatically reduced footprint. However, the volume expansion of mined material, the costs of double handling and transport of the material, the greenhouse and energy implications of transport, the remaining potential for generation of leachate and subsequent containment of seepage that may arise, and issues related to temporary storage, are all factors which can limit the application of this method. The back loading of ore haul trucks (where waste is loaded for the return journey) or the use of conveyers may provide solutions to the issue of double handling.

Text 2 Solid Backfill Mining Techniques
固废充填采矿技术

In the mining of coal resources, for every 1 t of coal mined, 10% to 15% (by mass) of solid waste rock is produced, mainly including waste rock discharged in tunnelling and washing. The general processing method thereof is to directly lift waste rocks onto the ground and accumulate them to form waste rock dumps. To solve a series of difficulties caused by discharge of waste rock, scholars proposed a method of backfilling goafs with coal mine solid wastes.

The key equipment for solid backfill mining includes backfill supports, a backfill conveyor and a self-shift loader. The self-designed backfill support used in this example was a Type ZZC10000/20/40, which includes a front top beam, back beam, six columns, four-bar linkage, compactor and support base (Fig. 15. 2). This support maintains the space for backfilling and mining so that backfilling and mining proceed in parallel. The compactor in the rear of the support consolidates waste rock in gob areas. The backfill conveyor (Fig. 15. 3) with unloading holes at the bottom hangs under the back beam of the support and can move along the back beam within a certain range (SGBC764/250). Opening and closing of the holes is controlled by telescopic jacks installed on the side. The self-shift loader transferring waste rock from the belt conveyor to the backfill conveyor is Type GSZZ-800/15. It includes a self-shift tail device, a belt stretching structure, an autonomous elevating structure, and a base sliding structure (Fig. 15. 4).

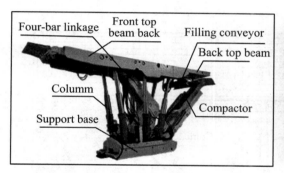 

Fig. 15. 2 Backfill hydraulic support structure   Fig. 15. 3 Working state of backfilling conveyor

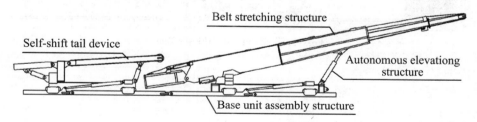

Fig. 15. 4 Self-shift loader

The solid backfill mining technology combines the mining and backfill processes, with the latter following the former. At the beginning, the backfill support advances after the coal is cut, then the backfill conveyor moves backwards along the support back beam. The backfill conveyor, self-shift loader and belt conveyor start in turn, and backfilling begins.

The backfilling process proceeds from the tail to the head of the backfill conveyor. When waste rock dropped from the unloading hole at the bottom of the backfill conveyor reaches a certain height, the next unloading hole is opened, and the compactor of the former support starts to tamp the waste. This process is repeated until the waste has been sufficiently tamped. Generally, two or three cycles are required. This first round of backfilling stops after the face has been fully filled. At this point, the first round of the backfill process is complete, and the backfill conveyor moves forwards along the support back beam.

The compactor functions to push waste rock that remains under the backfill conveyor towards the roof until the waste rock touches the roof and is fully compacted. The last process is to close all the unloading holes to backfill the space occupied by the conveyor's head section. After the first backfill cycle is complete, the backfill conveyor is pushed forwards to prepare for the next backfill process. The final backfilling effect is shown in Fig. 15.5.

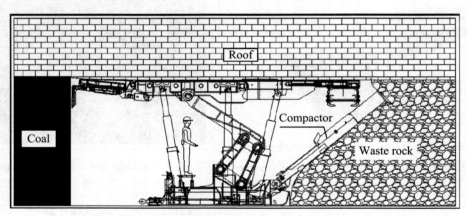

Fig. 15.5　Final backfilling effect

Unit 16　Future Mining Technology

未来采矿技术

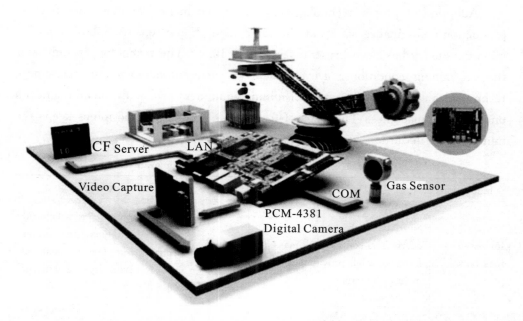

Fig. 16.1　Big data and interconnection technology of future coal mining

Text 1　A Strategic Approach for Sustainable Mining in Future　煤矿未来可持续发展策略

In order to meet the ever-increasing demands of the modern society, the mineral production is continuously increasing along with the scale of mining operations. However, mineral production is often not in consonance with conservation of environment and forests, which has several adverse impacts including air, water and soil pollution, socio-economic problems and effect on wildlife population and their behaviour. The technological developments in the field of underground mining, viz. mass production equipments, roof support, communication and automation are helping the decision makers to consider underground mining practice for sustainable mining while meeting environmental concerns.

Some of the measures that can be adopted for increased production from underground mines are:

1. Adoption of Appropriate Technology

Adoption of proper technology is one of the major parameter to increase production and enhance safety at the same time. Some selected innovations for mining industry has been presented in Table 16.1. The technology should suit the geo-mining conditions. Development of semi-autonomous and teleremote technologies enable the operation of mining equipments without human operators on-board the machines. This type of technology in underground mines keeps the miners out of hazardous environments.

Table 16.1 Selected recent mining industry innovations

Innovation	Description	Purpose	Outcomes
Improved rock bolts	New rock bolt design to absorb energy and control rock mass deformation while providing containment of materials.	Human safety	Improved control of rock failures, collapses of stopes and drifts; Reduced risk of rock bursts at increased mining depths and mining scales.
Collision avoidance system	Personnel and vehicle tags communicate wirelessly with moving vehicles; Driver alert and vehicle unit display of number of people and other vehicles in proximity.	Human safety	Vehicle operators are alerted to the presence of personnel or other vehicles in the vicinity.
Trapped miner location system/ paging system	Very Low Frequency (VLF) signal can penetrate through earth over large distances.	Human safety	Reliable means of quickly locating trapped miners; Reliable means of transmitting alert, warning and evacuation messages.

Continued

Innovation	Description	Purpose	Outcomes
Intelligent drill rigs	Intelligent drill rigs: - drill according to drill and bolt plan design; - ongoing measurement and reporting on tunnel profile; - remote access to program new drill plans; - ongoing monitoring of status of drill rigs and remote diagnostics.	Process/ efficiency improvements	Improved/more precise tunnel drilling; More rapid response/navigation of drill; Eliminates the need for surveyors; Enables remote evaluation of drill profile.
Extra low profile (XLP) mining equipment	Track mounted XLP dozer, bolter and drill rig capable of operating in a "narrow reef" (< 1.2 m height) and undulating mining environmentenvironment	Improved access/ efficiency improvement	Reduced waste rock/increased useful excavation; More accurate drilling and higher face advance.

2. Communication and Safety

Safety is a major concern for underground mining operation. However, there have been a number of developments for improvement of safety in underground mines. Many types of equipment are now fitted with onboard camera systems which monitor the working face locating personnel in remote locations and provide a real time look at the progress of the operation. Roof support in underground mining is a major challenge. As a result of this demand for increased production, manufacturers have made significant advances in development of advanced support systems. Several Russian and Australian mines have recently installed the latest continuous miner-bolter equipment(Fig. 16.2) which has significantly increased the reliability and development rates. The area of underground mine communications is undergoing rapid technological development and consequently several generations of communication systems are currently in use at mine sites. Radio-frequency identification (RFID) tagging systems are being used in mines to identify general location and information of tagged workers at the face. Wireless sensor networks have been developed for structural as well as gas monitoring. By regulating the mesh sensor network deployment and formulating a collaborative mechanism based on the regular beacon strategy, the network is able to rapidly detect structural variations caused by underground collapses. The collapse holes can be located and outlined and the detection accuracy is bounded.

Fig. 16.2　12 CM 30 miner bolter with bolting shield and roof bolts 2 m from head end

3. Automation

Automation is one approach for mines to improve safety and productivity but equipment availability is also key to every underground mining application. Unplanned downtime can slow down or stop any operation; however, equipment monitoring technology is helping operations track the health of their equipment. Proactive equipment management allows mine operators to review fault codes, track maintenance records and plan scheduled maintenance. In underground applications like room and pillar, this could mean the difference between a continuous miner in full production or a complete shutdown. Automated production drills have been available since the mid-1980s. Automated LHDs are now commercially available although automation of the dig cycle is problematic in anything but very well with broken rock. Automated trucks have operated reliably at the Finsch mine in South Africa for some time. Longwall coal mines have achieved partial automation of a relatively repetitive (continuous) mining system by automating one easily defined machine operation or task while the rest of the operations remained manual. The focus today has shifted to build the "autonomous mining system" that can carry out tasks automatically or with a

Unit 16 Future Mining Technology 未来采矿技术 · 159 ·

minimum of external control. Under full automation, a machine controls all aspects of its functions including monitoring and correcting for defects. These machines remove humans from hazardous areas, increase productivity as mining equipment moves faster, cover longer distances, and require fewer operators to control machines.

Text 2 Digital Mine 数字化矿山

Digital Mine virtual reality visualization technology is based on three dimensional visualization technology, data, warehouse technology, the full realization of deposit, geological modeling, reserve calculation, measurement data, fast mapping, underground mining system design and production simulation to a variety of engineering; work of the chart quickly generates a visual, digital and intelligent modern technology. For the simulation of underground production, thus further realize the importance of mine safety in production there.

Recent information technology advances have made possible the Digital Mine; in which information associated with all mining processes, however dispersed, would be integrated with support systems using new communication technology. The Digital Mine will provide the information infrustructure to serve as the foundation for mine wide automation.

1. Introduction

A digital mine (DM) is one where all interrelated communications are in digital format. Production, safety, geography, geology, construction and the like are linked by digital information systems. The essence of a DM is to use information and automation technologies along with modern control theories both to describe and control the mine operations systematically and dynamically. This enables highly effective safe green mining and ensures continuing improvement in mine economics. It also allows understanding and protection of the natural mine ecosystem. By now, the digital mine models proposed by others are too academic to manipulate. Given the demand of digital mines the construction of a practical three layer digital mine model is proposed by us. Two basic platforms, a uniform transmission network platform and a uniform data warehouse platform, are presented and discussed for the constructing of a DM.

2. Three Layer Digital Mine Model

For the realization of a DM we propose a three layer digital mine model (TLDMM) shown as Fig. 16.3. In Fig. 16.3 the lowest layer is the information collection and comm and dissemination layer. It collects the information from each production and safety subsystem and delivers the feedback control signals to the subsystems. All of the information and controlling signals are transmitted through a uniform transmission network. Although 14 subsystem modules are shown in Fig. 16.3, the number of modules could be increased or decreased according to the demand of a given mine. Most of the subsystem information in this layer is collected automatically; however it is acceptable that some of the subsystem information and data are entered manually. It is possible to integrate various production and safety subsystems into the control network transmission platform which then becomes a uniform, network based, collection and control system.

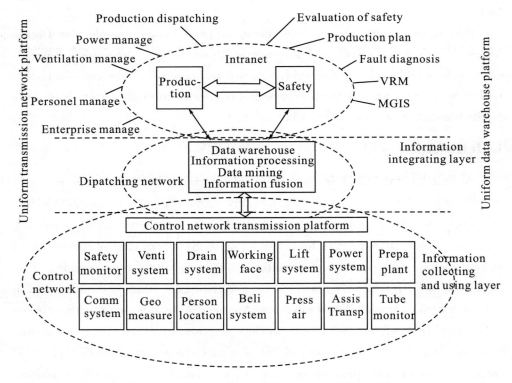

Fig. 16.3 Three layer digital mine model(TLDMM)

The layer in the middle is an information integrating layer. It is clear that each subsystem operates using its own data structure. For the DM mentioned above the data in the integrated automation system should have a standard form at and be processed in a standard way in order to enable seamless integration of subsystems. For example, when we want to evaluate the mine safety status online, the information we need comes from several different subsystems, such as the environment monitoring system, the ventilation control system, the mine pressure monitoring system, the underground water outburst and water level monitoring system.

Data warehouse technology and metadata techniques are the key technologies in the information integrating layer. The value of information coming from mine locales and their interrelated environments is enhanced in this layer.

The uppermost layer is a management and decision making layer. As we have mentioned, the value of information has been enhanced in the information integrating layer. The purpose of the enhancement is to allow use of the information to describe and control systematically and dynamically the management and safe production of the mine, to ensure continuous improvement of mine economy and to insure the stabilization of the natural mine ecosystem. The software modules listed in Fig. 16.3 are for this purpose. Just like the subsystem modules in the information collecting and dissemination layer, the number of software modules in the management and decision making layer could also increased or decreased according to the demands of a given mine.

Several aspects of the TLDMM shown as Fig. 16.3 are similar to other DM models we have mentioned. The characteristics of the TLDMM described as follow:

First, the TLDMM is not an unalterable model. Its scale could be adjusted by adding or removing subsystem and software modules given the factors of a scale of a particular mine, level of modernization and mining technology. Therefore, using this model to develop a DM is appropriate for various and different kinds of mine.

Second, it is apparent that there are two basic platforms in TLDMM. One is a uniform transmission network platform while the other is a uniform "data warehouse" platform, an operating database complete with its infrastructure. The two platforms are the foundation of a DM.

Third, the model crosses major fields. All professionals of these different subjects can play a role in the model.

Finally, the model is suitable for digitizing the entire mine. The technology may be introduced stepwise as the mine is developed or retrofitted to existing situations.

Text 3　Automation Mining Technology　无人采矿技术

1. Mining Automation Program

On January 1st 1996 Inco Limited, Tamrock OY (now Sandvik Tamrock), Dyno Industries AS (now Dyno Nobel AS) and the Canadian Centre for Mineral and Energy Technology (CANMET) formed a consortium for the development of robotic mining equipment called the "Mining Automation Program" or MAP. The program has been a-5-year research and development initiative that combined the talents of the above organizations to develop integrated systems of mining equipment and processes for tele-remote mining or Telemining (TM).

The MAP vision is: "To create a fully teleoperated mine including mine development, production drilling, loading and blasting, and mucking."Granted the vision of MAP is far-reaching but it addresses some very real and immediate challenges:

• The need for continued improvements to the safety and health of miners, who, despite progress and rigorous standards in this area, are always at some risk while underground.

• The need to improve the economics of hard rock mining in the face of mounting global competition.

• The need to increase the number and size of economically viable ore bodies. Many areas now running out of easily accessible, high-grade ore deposits. To meet the continued global demand for ore, there is a need to find ways to reduce costs so economic case can be made for developing new lower grade deposits or for mining at greater depths.

Telemining (TM) addresses these challenges in several ways. It greatly reduces the risk of accidents by removing operators from exposure to underground hazards and it provides for significant improvements in productivity and cost by maximizing manpower utilization and increasing quality & profitability.

Today, just to reach their work area, miners must travel thousands of feet underground in elevators and walk up to a mile through tunnels. With

Telemining (TM) operators are at the rock face and ready to start production soon as as they arrive at work. Telemining (TM) allows machines to enter an area immediately after blasting, without waiting for dust to clear or seismic activity to settle. It can greatly reduce the need for costly underground ventilation and cooling systems. And finally, it offers significant potential to multiply operator productivity by allowing them to operate several machines simultaneously.

2. Telemining (TM)

Telemining (TM) is used to describe a mining process that combines the use of remote operation (potentially with some pre-automation or autonomation), positioning & process engineering and monitoring & control.

The main technical ingredients of Telemining (TM) are:
- Advanced underground mobile computer networks.
- Underground positioning and navigation systems.
- Mining process monitoring and control sof tware systems.
- Mining methods designed specifically for Telemining (TM).
- Advanced mining equipment.

Reviewing the importance of each of these components, it was evident within MAP that the technical feasibility of the program hinged on two of the above.

Making them the cornerstones of the program. They were the successful development of a high capacity communication network, an accurate underground positioning and navigation system and finally the electronics and software to integrate all of this together. Development of the mining equipment that could perform all the mining functions was the next critical technology requirement along with explosive supplies that could be delivered to the hole teleremotely.

3. MAP Technology Elements

- Broadband Communications System. Remote operation requires operators to have their senses available (seeing, hearing, smelling, touching, and tasting) just as if they were in the mine.

The Broadband Communications System allows multiple pieces of mining equipment to be operated from the surface using mobile telephones, handheld computers, mobile computers on board machines and multiple video channels.

- Positioning & Navigation. To apply mobile robotics to mining, navigationand accurate positioning systems are an absolute necessity.

Inco has developed underground positioning systems based on inertial navigation technology and that have enough accuracy to locate the mobile equipment in realtime at the tolerances necessary for mining. Practical uses of such systems include machine set up, hole location and remote toping ultimately moving to surveying. The laser scanners provide data, which is imported into the computer system and used determine the machines location with the plan. This kind of positioning will allow the application of advanced manufacturing robotics typically applied to machines bolted to the floor to be used in mining; with mobile machines.

- Drills. The operator trams, sets up, monitors the drilling, and tears down from above ground using 4 video cameras mounted on the rig as his eyes. Information from the computer system located on surface directs the machine as to which angle, location and depth to drill the holes. Data collected from the drill consists of hole location, direction, depth monitoring, time, and hole movement, etc.

- Detonator and Explosive Delivery System. Dyno Nobel implemented 2 explosive systems in MAP. The first was prototype electronic detonator/initiation system (DynoRem EDI) and the second a bulk, repumpable, variable density, variable energy, emulsion explosive system.

The electronic detonator uses microchip technology to provide the capability for remote programming of accurate firing times (1 ms ~ 5000 ms) in 1 ms increments and remote computer aided initiation from the surface through the MAP broadband communications system. Unlike conventional electric or nonelectric detonators, where the time is built-in at the factory and unchangeable, the microchip electronic detonator's timing is addressable for programming by the blasting equipment and its timing is precise (1 ms).

The DynoRem ED blasting equipment is hard wired to the electronic detonator blasting circuit, and is remotely programmed with what electronic detonator chip address it is to communicate with and what time that address is planned to fire.

The tele-operated explosive loading machine has been one of the most challenging projects for Dyno Nobel in MAP. Through this project's development process, it has been learmned that many gaps in equipment and explosive's technology still stand between complete tele-operation and the needs for manual

Unit 16 Future Mining Technology 未来采矿技术

intervention in various parts of the explosive loading process. The most significant gaps being a way:

- To remotely find & identify the boreholes.
- To remotely clean the boreholes.
- To make up explosive primers and load them into the boreholes.
- To remotely connect the individual detonator legs wires or eliminate the need to do so.
- Load, Haul and Dump Machine (LHD). Tele-remote LHDs are the most proven automated machines in use at MAP. INCO has used this and similar types equipment for several years now in there operating mines and they have been responsible for moving millions of tons of ore.

4. Mining Operations System (MOS)

Mining Operations System (MOS) is a mining process monitoring system that captures information about the telemining process and makes it easily accessible to engineering, management, teleoperation, maintenance, and service personnel through the intra net.

With real time process information, informed decisions regarding each operation can be made. Plans can be reviewed. Work can be scheduled. Equipment can be located and their status reviewed. Maintenance and service can be scheduled. Process steps and continuous process improvements can be monitored.

5. MAP Technology Goals

MAP's goal is to shorten cycle times and lower costs while improving productivity and safety through bringing control of the entire mining cycle, including development and production, to a surface based operation center. The four partners of the MAP consortium are currently process testing telemining prototypes that cover mucking, drilling, loading and blasting and positioning. Development/Drifting Process, Mechanized Cut & Fill Production Process and Long-Hole Production Process testing.

Text 4　Virtual Reality (VR) Mining　虚拟采矿技术

1. Introduction

Virtual Reality (VR) is a continuously evolving new computer technology, which allows users to interact with computers in a new way. VR provides great opportunities for the minerals industry and is described by Aukstakalnis and Blatner as "a way for humans to visualize, manipulate and interact with computers and extremely complex data". VR systems are real-time computer simulations of the real world in which visual realism, object behaviour and user interaction are essential elements.

Virtual Reality is also a simulator, but instead of looking at a flat screen and operating a joystick, the user who experiences VR is surrounded by a three-dimensional computer generated representation, and is able to move around in the virtual world and sees it from different angles, to reach into it, grab it and reshape it. As the power of VR increases so too do its applications. VR has already been shown to be an effective tool in many industries. Surgeons may use VR to plan and map out complex surgeries in 3 dimensions, which allows them to view past the skin of the patient before a knife is even picked up. Realestate agents may use virtual reality to give clients a walkthrough of an estate, from the comfort of their own home.

Virtual Reality provides the best tools for accident reconstruction, training and hazard identification by immersing the trainee in an environment as close to the real world as possible. Through safety, visualization and education, VR promises many improvements for the minerals industry.

2. The Benefits of Using VR

Inadequate or insufficient training is often blamed for most of the mining fatalities. There is no doubt that the use of VR based training will reduce these injuries and fatality numbers. Justifying the use of VR in the minerals industry to improve safety is difficult to sell without hard evidence and quantified numbers. It is obvious that a considerable proportion of operating cost results from operators, or maintenance errors.

VR technology has a role to play in cost reduction in this area through improved planning and communications, and this is only one facet of the industry. VR multimedia training can dramatically reduce the cost of delivering training by decreasing learning time for trainees and instructors, the need for expensive and dedicated training equipment (physical mock-ups, labs, or extra equipment for training purposes), and travel expenses.

The difference between the conventional and VR training is that VR immerses trainees in realistic, functional simulations of workplace and equipment and it demonstrates mastery of skills through performance of tasks in multiple scenarios. Research Triangle Institute stated that maintenance mechanics in remote field locations who require training on expensive equipment which is unavailable for trainee practice had showed a 4 to 1 factor improvement using VR training. This translates into tremendous savings in labour and travel expenses.

3. Applications of Virtual Reality in the Mineral Industry

Recent advances in virtual reality and the increase in the power of modern computers has allowed for the rapid expansion of VR applications. It is estimated that over 60,000 commercial companies and approximately 3,600 educational institutions now use VR throughout the world. The mineral industry research and educational institutes in Australia have been more reluctant in embracing this technology compared to other countries and industries. The Institute of Sustainable Minerals Industry, Advanced Computer Graphics and Virtual Reality Research Group (SMI-VR) at the University of Queensland, has recently been established to promote this evolving technology and develop VR applications for the minerals industry. The SMI-VR research group has developed a number of successful VR applications for data visualization, accident reconstructions, simulation applications, risk analysis, hazard awareness applications and training (training of drivers, and operators).

The following are just some of the VR applications developed or under development by the SMI-VR research group:
- Drill rig training simulation.
- Open pit simulation (Fig. 16.6 and Fig. 16.7).
- Underground hazard identification and barring down training simulation.
- Instron rock testing simulation.
- Accident reconstruction.
- A library of 3 dimensional mining equipment.

- Ventilation survey and real time monitoring simulations.
- Virtual mining methods.

Fig. 16.6　Loading a truck: cabin view　　　Fig. 16.7　Haul truck training simulator setup

4. Drill Rig Training Simulation

Training is becoming a high priority for the mineral industry due to high injury and fatality rate. Although there is no substitute for real world training, VR offers the necessary tools to reduce the cost of training and improve safety. Through the use of VR training, personnel can learn off site without disturbing production schedules or interfering and endangering expensive machinery with untrained personnel. Big TED virtual drill rig system developed by SMI-VR group is a good example of such training systems. Big TED has fully functional control systems to operate the rig with visual warnings and 3-dimensional sound effects. The simulation is made to emulate, as closely as possible, the actual lab drill.

In the operator training mode, the user is allowed to operate the rig in the same manner as in real life. The operation of the drill rig involves turning on the power, air and water supply in the right order and carefully applying rotation and thrust to the drill head without damaging it. The machine responds to any mistakes made by the operator with 3-dimensional sound effects and pop-up text messages.

5. Underground Hazard Identification and Barring down Training Simulation

Virtual reality can be used to train individuals to perform tasks in dangerous situations and hostile environments, such as in an underground mine accident or toxic gas environment. In addition to the assurance of safety, the use of a virtual training of the trainee's environment gives the trainer total control over many aspects of performance. The virtual environment can be readily modified, either to provide new challenges through adjusting levels of difficulty or to provide training prompts to facilitate learning.

6. Virtual Mining Methods

Virtual reality is increasingly being used as an educational tool. There are increasing demands today for ways and means to teach and train individuals without actually subjecting them to the hazards of particular simulations. VR is an emerging computer technology, which has strong potential to overcome a number of limitations of conventional teaching methods. The most important benefits of using VR in education are that it is highly immersive, interactive, visually oriented, highly sensory, colourful and generally exciting and fun.

The use of VR in mining education has a number of benefits. For example, underground developments can be shown through a transparent layer of rocks to visualize. This gives students a better understanding of how the developments are constructed and their orientations in respect to the orebody.

Students can be taken to a mine site and be given a tour of the mine in an immersive and realistic environment to introduce them to the operations and hazards. VR techniques can also be used to teach students how:

- to operate a machine.
- a truck — shovel simulation works.
- to carry out dangerous operations such as blast tie in and blasting exercise.
- to carry out a laboratory test i.e. Uniaxial Compressive Strength.
- a mine is developed and a mining method is utilized, etc.

Terms 常见表达

虚拟现实 virtual reality
虚拟环境 virtual environment
遥操作系统 teleoperator system
增强现实 augmented reality
合成环境系统 synthetic environment system

共用控制器 shared controller
共用显示器 shared display
分散型控制系统 distributed control system
系统结构 system architecture
过程控制级 process control level
监控级 supervision level

管理级 management level
直接数字控制站 direct digital control station
数据采集站 data acquisition station
操作员站 operator's station
工程师工作站 engineer's work station
多路转换器 multiplexer
可编程序控制器 Programmable Logic Controller(PLC)
非对称数字用户线 Asymmetrical Digital Subscriber Loop(ADSL)
异步传输模式 Asynchronous Transfer Mode(ATM)
计算机辅助设计 Computer Aided Design (CAD)
计算机辅助制造 Computer Aided Manufacture(CAM)
CA认证 Certificate Authority(CA)
现在集成制造系统 Contemporary Integrated Manufacturing Systems (CIMS)
存储器直接访问 Direct Memory Access (DMA)
高清晰度电视 High Definition Television (HDTV)
IP电话 Internet Protocol Telephone (IP)
安全电子交易 Secure Electronic Transaction(SET)
安全软插座层协议 Secure Sockets Layer Protocol(SSLP)
机顶盒 Set Top Box(STB)
TCP/IP协议 Transmission Control Protocol/Internet Protocol
波分复用 wavelength-division multiplexing
带宽 bandwidth
地理信息系统 geographic information system
电路交换 circuit switching
电信网 telecommunication network
电子商务 electronic commerce
电子数据交换 Electronic Data Interchange(EDI)
分组交换 packet switching
光传送系统 optic transmission system
光纤通信 fiber-optic communication
国家信息基础结构 National Information Infrastructure(NIF)
集成软件 integrated software
集成电路 integrated circuit
全球信息基础设施 global information infrastructure
人机界面 man-machine interface
软件包 software package
生物芯片 biochip
数据库服务器 database server
数据速率 data rate
数据通信 data communication
数据压缩 data compression
数字电视 digital TV
同步网 synchronizing network
图像通信 image communication
微波通信 microwave communication
卫星通信 satellite communication
无线电通信 radio communication
无线应用协议 wireless application protocol
无纸办公室 paperless office
系统集成 system integration
虚拟专用网络 virtual private network
智能网 intelligent network
专家系统 expert system
综合业务数字网 integrated service digital network

Appendix A English Writing of Science and Technology

科技英语写作

A.1 Introduction 概论

科技论文是科学研究成果的一种展示,指作者以论文的形式向外界公布自己的科学研究成果。科技论文的撰写与发表是科学研究中一项极其重要、不可马虎的工作,对于非英语母语的作者来说,如何用英文准确、清晰地表达研究内容与获得的成果,如何使论文顺利发表,成为其必须面对的问题。本附录详细介绍了论文写作与发表过程中的一系列具体工作,以期对读者今后的写作提供一些有益的帮助。

A.1.1 英语科技论文的主要结构组成

英语科技论文的整体组织结构与语言风格规范尤为重要,和建造楼房一样,写一篇论文也需要一份蓝图。英语科技论文最常见的是"IMRAD"结构,即:Introduction, Methods, Results and Discussions。

IMRAD结构的论文简单、清晰明了且逻辑性强,因而这一结构被近代学者广泛采用。这种结构的论文,首先阐述研究的课题与目的,有了清楚的定义以后,再描述研究的方法、试验手段和材料,最后对结果和结论进行详细讨论。作者的任务就是通过 IMRAD 结构,将上述信息介绍给读者。

除了文章的结构以外,语言也是非常重要的因素。科技英语与通常的文学语言有明显的不同,它的特点在于客观、直叙、简练、准确,所叙述的过程具有很强的可操作性。它有两个重要原则:

(1) ABC 原则:Accuracy(准确),Brevity(简洁),Clarity(清晰)。

(2) The Three C's 原则:Correct(正确),Concise(简洁),Clear(清楚)。

文学和艺术都有较强的娱乐性,目的是愉悦读者。所以文艺作品使用较多的形容词、副词及一些没有清楚定义的词。例如用 enormous、immense、tremendous、huge、large 来形容大,用 tiny、small、little、petit、light 来形容小。一些政客也一样,他们不用准确的语言,喜欢用 acceptable、reasonable、satisfying 来形容结果,也常用一些不确定、模棱两可的词如 perhaps、maybe 来评论事情。这种风格及这类词汇在科技论文里一般是不能使用的,科技论文需要的是清楚、准确、尽可能量化的表达方式。

对于我们来说,第一步需要写出语法正确的句子(Correct Sentences),不少人使用中国式的表达方式写英语句子,造成词法或语法不通,我们称之为"Chinglish"(Chinese English)。投到国外杂志的科技论文常常被退回来修改,国外评阅人的意见是"What you described in the paper is OK. But your English is poor. Please improve it. (or Please rewrite your paper entirely)"其问题在于很多人对英语科技论文的特点和常用句型不了解。第二步是要努力写出地道的句子(Standard Sentences)。在科技英语中,每个专业都有自己的一些特定词语及固定表示法。多读一些本专业的英语书刊可增强语感,培养用英语思维写作的能力,注意一些特定的用法并将其用于自己的论文中,这样能够不断提高自己的科技英语写作能力。

下面对英语科技论文的主要结构做简要介绍。一般来说,一篇完整规范的学术论文由以下各部分构成:

(1) Title(标题)　　　　　　　(2) Author and Contact Information
　　　　　　　　　　　　　　　　　　(作者和联系方式)
(3) Abstract(摘要)　　　　　　(4) Key Words(关键词)
(5) Table of Contents(目录)　　(6) Nomenclature(术语表)
(7) Introduction(引言)　　　　 (8) Methods(方法)
(9) Results(结果)　　　　　　　(10) Discussions(讨论)
(11) Conclusions(结论)　　　　 (12) Acknowledgement(致谢)
(13) References(参考文献)　　　(14) Appendix(附录)

其中(1)~(4)、(7)~(11)和(13)是必不可少的,其他几项根据具体需要而定。

A.1.2　科技论文的分类及特点

A.1.2.1　分类

科技论文的分类可按写作目的进行分类,也可按论文内容进行分类。

1. 按写作目的分类

科技论文根据写作目的可分为期刊论文、学位论文和会议论文。

(1) 期刊论文可在学术期刊发表,它在科技论文中占有最大比例。不同级别的期刊对论文的深度、广度要求虽然不同,但对论文的创新性都具有较高要求,期刊编辑部希望科技论文在选题和内容方面能够提出新的观点、新的发现、新的经验，要求在理论上具有较强的指导意义或在应用中有较高的实用价值。绝大部分期刊对论文的字数提出了限制，大致为 6000～8000 字。

（2）学位论文分为学士论文、硕士论文和博士论文，其中博士论文内容最多，字数通常也最多，硕士论文次之，学士论文字数最少。期刊论文一般只有一个主要内容，学位论文则可以有几个科技创新点，特别是博士论文，一般要求至少有四个创新点。

（3）会议论文供参加学术会议用。各种会议论文水平相差很大，其深度、广度通常有限，但会议论文的特点是可以更快地反映当前研究的现状，参加者或读者可由此了解国内外的研究热点及研究水平。

2. 按论文内容分类

科技论文根据内容可分为实验型、理论型和综述型三种。

（1）实验型论文的重点在于通过科学实验提供事实，其中包括新方法或新工艺条件，这种论文常有大量数据，同时包括由数据得出的一些规律。

（2）理论型论文可以完全不涉及实验，而是引用文献的实验结果，由此提出新规律、新模型、新计算方法，例如数学类论文。在同一篇论文中也常见实验和理论部分同时存在。

（3）综述型论文是对某一课题前期工作的评述，总结某一方法、技术。这种论文常常产生于文献调查后，对同方向的研究者有指导作用，是其今后工作的基础。综述型论文不但应包括对此课题前期工作的介绍，更重要的是应对这些文献的成果做出客观中肯的评价，指出优缺点，还应该指出今后的研究方向。

A.1.2.2 特点

科技论文特别是期刊论文，要有鲜明的论点、充分的论据、严密的论证以及确切的论断，因此作者撰写的科技论文一般应具有科学性、创新性、逻辑性、规范性和简洁性。

（1）科学性是科技论文的前提，即研究的对象应属于科技范畴，论述的内容要真实可靠，要以可靠的数据、真实的现象和已有的理论作为依据。

（2）创新性是科技论文最重要的条件，论文所揭示的事物的现象、属性、特征、运动规律及规律的运用，都力求是前所未有的、首创的或部分首创的，而不是重复、模仿、抄袭他人的工作。

（3）逻辑性指科技论文必须结构严谨、层次分明、前提完备、概念确切、推理严密、运算无误、分析透彻、判断准确。

（4）规范性指科技论文的格式、插图表格、计算单位、数学公式、数字用法、参考文献等，都必须符合有关标准与规范的规定。

（5）简洁性指科技论文须文字简洁、语句简练，论文中引用他人的文献只需摘录其主要观点和重要数据，注明参考文献出处即可。基础知识、复杂运算的中间步骤、计算机源程序等均可省略。

A.1.3 英语科技论文的文体特点

A.1.3.1 文体总貌

科技论文内容一般比较专业,语言文字正规、严谨,论文的结构已成格式化。科技论文一般不使用带有个人情感色彩的词句,而是以冷静客观的风格陈述事实和揭示规律。总的来说,英语科技论文有两大显著特征:

(1) 文体正式

科技论文用词准确、语气正式、语言规范,避免口语化的用词,不用或少用 I、we、our lab 或 you 等第一、二人称的代词,行文严谨简练,不掺杂个人的主观意识。

(2) 高度的专业性

科技论文均有一个专业范围,其读者均是各专业的科技人员。专业术语是构成科技论文的语言基础,其语义具有严谨性、单一性等特点。使用术语可使论文的意义准确、行文简洁。

A.1.3.2 词汇特点

英语科技论文的用词有别于其他类型的文体,其词汇有以下特征:

(1) 纯科技词

主要指那些仅用于某个学科或某个专业的词汇或术语。不同的专业有不同的专业技术词汇。随着科技的发展,新的学科和领域不断涌现,新的发明和创造层出不穷,新的专业术语也随之产生。每个学科和专业都有一系列专业术语,如 hydroxide、anode、isotope 等。

(2) 通用科技词汇

指不同专业都经常使用的词汇,但在不同的专业内却有不同的含义,如 transmission 在无线电工程学中表示"发射、播送",在机械专业中则表示"传动、变速"的意义,而在医学中表示"遗传"。又如 operation 一般理解为"操作",在计算机科学中指"运算",在医学中则指"手术"。这类词用法灵活,适用范围广泛,使用时要慎重。

(3) 派生词

英语的构词法主要有合成、转化和派生三大手段。这三种手段在科技论文的词汇构成中都有大量的运用,其中派生词,也就是加前、后缀构成的不同词出现频率远远高于其他方法构成的词,如 antibiotics、microbiology、macrocode 等。

(4) 语义确切、语体正式的词

英语中有大量的词组动词,通常由动词+副词或介词构成。这类动词意义灵活,使用方便,但是其意义不易确定,所以在科技论文中不宜过多使用,而宜采用与之对应的、意义明确的单个词构成的动词,如通常使用 absorb 来替代 take in,用

discover 来替代 find out，用 accelerate 来替代 speed up。这类动词除了具有意义明确、精练的特点外，还具有语体庄重、正式的特点。下面是一组对照词语，右侧的表达方式更适用于专业论文或书面文体。

use up —— exhaust	push into —— insert
throw back —— reflect	put in —— add
carry out —— perform	think about —— consider
get rid of —— eliminate	take away —— remove
get together —— concentrate	fill up —— occupy
keep up —— maintain	hang up —— suspend

英语词汇有两套系统：一套来自英语中的本族词，这一类词中单音节词居多，语体口语化或中性；另一套系统源于法语、拉丁语，以多音节为主要特点，语体正式。科技论文多使用源于法语、拉丁语和希腊语的词汇。下列两组词中，第二列为正式文体所常用的词：

非正式文体	科技论文或正式文体
finish	complete
underwater	submarine
buy	purchase
enough	sufficient
similar	identical
handbook	manual
careful	cautious
help	assist
try	attempt
get	obtain
about	approximately
use	utilize

(5) 合成词

合成词表示直接，能使语言简练，所以在论文中出现的频率较高，如 horsepower、blueprint、waterproof、wavelength 等。

(6) 缩写词

有些专业术语由词组中首字母构成，这在科技论文中经常使用，如 CAD(Computer Assisted Design)、OOPL(Object-oriented Programming Language)、OOA/D(Object-oriented Analysis Design)等。

A.1.3.3 句法特点

(1) 较多使用被动结构

科技论文叙述的是客观事物、现象或过程,而使其发生的主体往往是从事某项工作的人或装置,使用被动结构能使论文叙述更显客观。此外,把描述和研究的对象放在句子的主语位置上也能够使其成为句子的焦点,让读者将注意力集中在叙述的事物或过程上。

(2) 较多使用动词非谓语形式

动词非谓语形式是指分词、动词不定式和动名词。科技论文常使用非谓语动词形式有两个主要原因:

(a) 非谓语动词形式能使语言结构紧凑,行文简练,例如:

Numerical control machines are most useful when quantities of products to be produced are low or medium; the tape containing the information required to produce the part can be stored, reused or modified when required.

其中的 the tape containing the information 和 required 两个分词短语,如果不用非谓语形式,只能分别用 the tape which contains the information ... 和 the information which is required to produce... 等从句的形式表达,导致语句冗长。

(b) 非谓语动词形式能够体现或区分出句中信息的重要程度,例如:

A theoretical framework is provided, consisting of negative reinforcing feedback loops that act as drivers behind future industry change.

在句中动词 is provided 表达了主要信息,现在分词短语 consisting of ... 提供细节,即非重要信息。

(3) 倾向于多用动词的现在时

科技论文多用一般现在时来表述经常发生的无时限的自然现象、过程等。例如:

The purpose of this technology is to detect driver drowsiness.

(4) 长句多,句子结构较复杂

科技论文一般描述较复杂的活动和关系,所以经常使用结构严谨的长句和复杂句,经常出现多个分句并列或复合句中从句套从句的现象。

A.2 Standards and Skills of Language 语言表达的规范与技巧

科技论文的英语表达在措辞和句法结构方面有一些基本的规范与技巧,作者在撰写论文时应该了解和掌握。

A.2.1 如何使表达准确

A.2.1.1 选词要正确

科技信息重视精确性、准确性和具体性,在措词时应力求精确,要使用确切具体的词,否则会影响论文的质量。例如:

Our measurements are not precise because the experimental apparatus was in poor condition.

Our measurements are not precise because the scale was not working properly.

在选词时还要考虑其意义的内涵差别,如 equipment、installation、instrument 就因其内涵不同在用法上有差异。equipment 是不可数名词,意义上往往指有形的、看得见的硬件设备;installation 也指设备、设施,但一般是指安装或装配上去的设备,不可以随时拿走或拆卸;而 instrument 常指精密度较高的器械、仪器、仪表,例如:

The marines took with them full combat equipment including tanks, artillery, jeeps, and flame throwers.

This laboratory does not even have a heating installation.

This model has separate instruments for oil pressure, water temperature, fuels gauge, and ammeter.

A.2.1.2 细节描述的程度要适当

科技论文既要有概括的表达,也要提供细节,以说明和支持概括的表述。例如:

Our new process reduces emissions of nitrogen oxides from diesel engines and industrial furnaces.

这样的语言太笼统,只有添加细节,才能让读者了解新的工作程序的重要性,例如:

Our new process eliminates 99% of nitrogen oxide emissions from diesel engines and industrial furnaces. Previous processes have, at best, reduced nitrogen oxide emissions by only 70%.

细节要适当,要有选择。选择的原则是重点突出,对于公知公认的或同领域研究者熟知的内容不需要细化,可适当删减。添加细节要能够使读者获得新信息。

A.2.2 语义要连贯

科技论文是一篇语义连贯的文章,构成文章的句子之间必然存在某种逻辑上的联系,语句的衔接和连贯是语句成篇的重要保证。实现衔接与连贯主要有以下几种方式：

A.2.2.1　使用连接词

英语中使用连接词将句子衔接起来，以使意义连贯。句子之间缺乏必要的衔接手段，就显得非常松散。例如：

① Gold, a precious metal, is prized for important characteristics. ② Gold has a lustrous(有光泽的)beauty that is resistant to corrosion. ③ It is suitable for jewelry, coins and ornamental(装饰性的) purposes. ④ Gold never needs to be polished, and will remain beautiful forever. ⑤ A Macedonian(马其顿的) coin remains as untarnished(失去了光泽的)today as the day it was minted(铸造) twenty-three centuries ago. ⑥ Important characteristics of gold is its usefulness to industry and science. ⑦ It has been used in hundreds of industrial applications. ⑧ The most recent use of gold is in astronauts' suits. Astronauts wear goldplated heat shields for protection outside the spaceship. ⑨ Gold is treasured for its beauty, for its utility.

将该段落修改后如下：

① Gold, a precious metal, is prized for two important characteristics. ② First of all, gold has a lustrous beauty that is resistant to corrosion. ③ Therefore, it is suitable for jewelry, coins and ornamental purposes. ④ Gold never needs to be polished, and will remain beautiful forever. ⑤ For example, a Macedonian coin remains as untarnished today as the day it was minted twenty-three centuries ago. ⑥ Another important characteristics of gold is its usefulness to industry and science. ⑦ For many years, it has been used in hundreds of industrial applications. ⑧ The most recent use of gold is in astronauts' suits. Astronauts wear goldplated heat shields for protection outside the spaceship. ⑨ In conclusion, gold is treasured for its beauty, for its utility.

在添加了必要的连接词后，段落变得流畅连贯了。

请再看一例：

① Lasers have found widespread application in medicine. ② For example, they play an important role in the treatment of eye disease and the prevention of blindness. ③ The eye is ideally suited for laser surgery because most of the eye tissue is transparent. ④ Because of this transparency, the frequency and focus of the laser beam can be adjusted according to the absorption of the tissue so that the beam "cuts" inside the eye with minimal damage to the surrounding tissue — even the tissue between the laser and the incision(切口). ⑤ Lasers are also more effective than other methods in treating some causes of blindness. ⑥ However, the interaction between laser light and eye tissue is not fully understood.

从上文可以看出，连接词的使用在保证语句连贯方面起着举足轻重的作用。

科技论文中常用的连接词可大致分为以下7类，表示不同类型的句间关系：

(1) 表示时间、顺序关系：after，before，first，second，then，next，finally，later，meanwhile(其间)，in the past 等。

(2) 表示空间关系：here，there，nearby，under，below，in front of，in the middle of，at the back of 等。

(3) 表示附加、递增关系：and，also，again，in addition，besides，furthermore(此外，而且)，moreover，what is more(更甚者)等。

(4) 表示异同关系：however，nevertheless，but，although，yet，while，on the contrary，in spite of，instead，but，unlike，similarly，likewise，in the same way，like 等。

(5) 表示因果关系：since，because，as，result from，due to，thanks to(由于)，for this reason，therefore，consequently，accordingly，thus 等。

(6) 表示列举：for example，for instance，such as，namely(即，也就是)，in other words，that is to say 等。

(7) 表示总结：in conclusion，to sum up，in short，in a word，in brief，on the whole 等。

A.2.2.2 重复关键词

关键词是在科技论文中表达内容的重要概念，可能会在论文中被反复提及，重复关键词可以使语句连接紧密。例如：

The CPU memory unit is commonly called the internal memory of acomputing system, on older machines this memory usually consisted of magnetic cores ... Most computing systems also incorporate components that serve as auxiliary or external memory.

在上例中 memory 这一关键词重复4次，既使语句意义连贯，又有效地突出了主题。

A.2.2.3 使用主从结构

在论文中，句与句之间表达的意义关系并不是平行和并重的，有时它们之间有主次、轻重之分，有的表示背景与主题之间的关系，有的表示条件与结果的关系。在写作时，要用主从复合结构将这种意义的层次体现出来，否则，就影响意义的连贯。例如：

Some materials are liable to absorb moisture. This will adversely affect their insulating(绝缘) properties.

上例中的两个句子意义相关联。但是，由两个单独的句子表达，结构有些松

散,所以将两句合并,用主从复合结构表达,这样意义更加连贯,例如:

Some materials are liable to absorb moisture, which will adversely affect their insulating properties.

A.2.2.4 应避免错误

书写时要使句子保持连贯,避免犯下列错误。

(1) 误置修饰语的错误

修饰语(即定语或状语)应靠近所修饰的词。如果远离所修饰的词就成为误置修饰语,可能会影响句子的连贯性。

例1:The engineering student made a big mess wearing the red hat. （错误）

　　　The engineering student wearing the red hat made a big mess. （正确）

注意修饰性从句同被修饰成分保持明显的关系。

例2:She borrowed a car from her friend that was bad. （意义不清楚）

　　　From her friend, she borrowed a car that was bad. （正确）

要避免不一致的修饰语。不一致的修饰语是模棱两可的,既可修饰这个词又可以修饰那个词,使读者茫然。

例3:His doctor told him frequently to exercise. （模棱两可）

　　　His doctor frequently told him to exercise. （正确）

　　　His doctor told him to exercise frequently. （正确）

(2) 平行结构错误

平行结构就是句中表达同样意思的并列的两部分或更多的部分以同样的语法形式连贯地表现出来。换句话说,就是主语和主语、谓语动词和谓语动词、宾语和宾语、补足语和补足语平行,或名词与名词、形容词与形容词、副词与副词、从句与从句等平行。平行结构一般由相关连词 and,or,either ... or ... ,neither ... nor ... ,not only ... but also ... 等引导。当并列部分的语法形式不同时,就出现了平行结构错误。

例4:To design a device and making it are two different jobs. （错误）

　　　Designing a device and making it are two different jobs. （正确）

例5:Galileo found it difficult to believe that the sun rotates around the earth and the earth to be the centre of the universe. （错误）

　　　Gelileo found it difficult to believe that the sun rotates around the earth and that the earth is the centre of the universe. （正确）

(3) 模糊指代错误

模糊指代错误是指代词可指代两个不同的先行词。为了消除模糊指代可用名词来代替代词或阐明先行词。

例6:Our responsibility in the laboratory was to remove the labels from the dishes and wash them. （错误）

Our responsibility in the laboratory was to remove the labels from the dishes and wash the dishes. （正确）

(4) 泛指代的错误

泛指代中的先行词是一个泛泛的概念，是整个意思而不是一个特定的名词或名词结构。当代词指代两个以上的意思时，就会出现泛指代错误。this, that 和 which 的含糊使用常导致这种错误。

例7：He spent his time getting help with his income tax forms, which his wife considered unfair. （错误）

His wife considered it unfair that he spent his time getting help with his income tax forms. （正确）

A.2.3 英语科技论文常用句型结构

在写英语科技论文时，作者有必要熟悉和掌握一些重要的常用句型，进行记忆和模仿，这对提高英语论文的语言质量非常有帮助。以下句型按功能分类列出。

(1) 定义

定义是对一事物的本质特征或一个概念的内涵和外延的确切而简要的说明，常见的表达方式和句型有：

The term macromolecule $\begin{cases} \text{means } ... \\ \text{signifies } ... \\ \text{is considered to be } ... \\ \text{is taken to be } ... \\ \text{refers to } ... \end{cases}$

$\left.\begin{array}{l}\text{In this article}\\ \text{In this paper}\\ \text{In this essay}\\ \text{In this context}\\ \text{In computer science}\\ \text{For this purpose}\end{array}\right\}$ network $\begin{cases} \text{will be taken to mean } ... \\ \text{will be used in the } ... \\ \text{will be considered to be } ... \\ \text{will refer to } ... \\ ... \end{cases}$

例：Paradigm(范例) refers to models of inquiry that guides scientific work.

(2) 分类

在科技论文中，作者有时需要对描述、说明的对象按其性质、大小、作用等分成若干种类，然后依次阐述。常见的表达分类的句型有：

$$\left.\begin{array}{l}\text{break down}\\ \text{divide}\\ \text{subdivide}\\ \text{be classified}\end{array}\right\} + \text{...（总体）} + \text{into/as ...（部分）}$$

具体用法举例如下：

Classification breaks down the general topic into its related parts in a logical way.

This classification system divided carbides into two distinct categories.

Substances can/may be classified as gases, liquids and solids.

$$\text{The models}\left\{\begin{array}{l}\text{fall into}\\ \text{can be divided into}\\ \text{may be classified into}\\ \text{are of}\end{array}\right\}\text{two}\left\{\begin{array}{l}\text{major}\\ \text{general}\\ \text{broad}\end{array}\right\}\left\{\begin{array}{l}\text{groups.}\\ \text{classes.}\\ \text{categories.}\end{array}\right.$$

$$\text{There are two}\left\{\begin{array}{l}\text{major}\\ \text{general}\\ \text{broad}\end{array}\right\}\left\{\begin{array}{l}\text{groups}\\ \text{classes of ...}\\ \text{categories}\end{array}\right.$$

（3）比例关系

常见表达比例关系的句型有：

Most objects are big in proportion to the size of an atom but small in proportion to the size of the sun.

The ratio between students of science to students of engineering is 3∶2.

Pressure is inversely proportional to volume.

Air and fuel are mixed in a proportion of 15 to 1.

Three colors：red，green and blue will，if they are mixed together in the right proportion，give us white light.

In the final printed version，graphs and diagrams are usually reduced in scale by a factor of two or three.

A good coolant is formed by mixing water and soluble oil in a ratio of 10∶1 to 20∶1.

The acceleration of a body is directly proportional to the force acting on it.

（4）比较

常见表达比较的短语有 as a comparison, by comparison, by contrast, in contrast, by way of contrast 等,例如：

The arithmetic-logical unit is also responsible for choosing and comparing the appropriate information with a program.

By comparison, very little is known about the internal structure of the earth.

（5）举例

常见表达举例的句型有：

This can be illustrate by ...

A specific case can be provided to show ...

An example of this involves ...

This can be demonstrated through ...

Another typical example is ...

One example will suffice ...

（6）与……一致

A be $\begin{cases} \text{compatible} \\ \text{in conformity} \\ \text{consistent} \\ \text{in good agreement with B} \\ \text{in accord} \\ \text{in line} \\ \text{uniform} \end{cases}$

表达 A 与 B 一致还可用以下句型：

The meal contents found from this agree quite well with those valuated by a microscopic(精确的) method.

Results show that calculations are in good agreement with experimental data.

The calculated components may not match exactly the original design of the oscillator.

A.3　Contents of Science and Technology Paper 英语科技论文内容

科技论文是科学研究成果的书面表达形式，如何准确、清晰地表达研究内容和获得的成果，提高论文的可读性，合理安排论文的结构，使每一部份内容详实、论述完整、结构逻辑严谨，成为研究工作中一个不可忽视的重要组成部分。

前面已经对英语科技论文的主要结构做了简要介绍。下面对其主要部分做简要介绍。

（1）标题：科技论文出版之后，被人看到最多的是论文的标题。在期刊目录与互联网上，浏览论文标题的人可能会比最终阅读全文的人要多千万倍。因此，论文标题能否吸引大批潜在的读者,对提高论文的影响力至关重要。选择标题的要领在于用最少的词把最核心的内容表述出来。

(2) 作者和联系方式:论文的署名表明作者享有著作权且文责自负,同时作为文献资料,也便于日后他人索引和查阅。论文署名还便于作者与同行或读者进行研讨和联系,因此有必要提供作者的联系方式。

(3) 摘要:摘要是论文的缩影,是对论文的简单描述。摘要的作用是为读者提供关于文献内容的足够信息,即论文所包含的主要概念和所讨论的主要问题,使读者从摘要中可获得作者的主要研究活动、研究方法和主要结果及结论。摘要可以帮助读者判断此论文是否有助于自己的研究工作,是否有必要获取全文。一篇好的摘要应该具备以下要素:① 完整性——它是完全独立的,论文中的基本信息和要点都应该出现在摘要里。② 可读性——以通俗易懂的语言来描述复杂的概念和高深的问题。③ 科学性——使用标准、精练的词汇和语言,清晰紧凑地概述客观事实。④ 逻辑性——摘要整体结构严谨,思路清楚,基本素材组织合理。

(4) 前言:前言的作用是提供足够的研究工作背景和内容,通常包括以下信息:① 为什么写这篇论文,要解决什么问题。② 与课题相关的历史回顾,本课题在学科领域中所占的地位及课题的意义和价值。③ 研究所涉及的界限、规模和范围。④ 理论依据和实验设备基础。⑤ 预期目标。⑥ 概念和术语的定义。上述内容不必逐一介绍,可视具体情况进行取舍,应突出重点,关键是从一开始就吸引住读者的注意力,使读者了解为何选择这个课题以及这项研究为何重要。

(5) 方法:撰写方法部分时总体要把握重点突出、详略得当。对公知、公用的方法写明其方法名称即可;引用他人的方法、标准,已有应用而尚未为人们熟悉的新方法等应注明文献出处,并对其方法做简要介绍;对改进或创新部分应详细介绍。总之,实验技术和方法的介绍,既便于其他研究者可重复研究,又可以提高读者对该研究设计及其结构可靠性的信任程度。

(6) 结果和讨论:实验结果是对研究中所发现的重要现象的归纳,论文的讨论由此引发,对问题的判断推理由此导出,全文的一切结论由此得到,这一部分是论文的核心。在介绍实验结果时,作者要指明研究结果在哪些图表公式中给出,还应注意对结果进行说明、解释,并与模型或他人结果进行比较。结果常与讨论合并在一起。讨论是论文中的重要部分,在全文中除摘要和结论部分外受关注度最高,是读者最感兴趣的部分,也是比较难写的部分。在这部分中,作者要回答引言中所提的问题,评估研究结果所蕴含的意义,用结果去论证所提问题的答案。

(7) 结论:科技论文的结论部分与引言是相呼应的,是论文中继摘要、引言之后,第三次重申问题的重要性和研究的价值(问题的重要性分别在引言、摘要和结论中共重复强调三次)。读者往往在看过摘要之后,紧接着就看结论部分,以了解研究工作的主要成果,再决定是否有必要认真阅读全文或其中一部分。结论是论文实质内容的浓缩,论文作者应十分重视结论的写作。

A.3.1 标题

科技论文的标题是论文内容的高度概括,拟定标题时需要注意以下几点:

(1) 长度不宜过长,大多为 8~12 个单词(中文一般要求不超过 20 个字)。国际标准化组织规定,每条标题不要超过 12 个词,且除通用的缩写词或特殊符号外,标题内不使用缩写词和特殊符号。如果 12 个词不足以概括全文的内容,可使用副标题加以补充说明。

(2) 标题的用词:① 为了便于检索,标题中尽量使用能表达全文内容的关键词,即有多个文章的关键词出现在标题中。② 标题通常由名词短语构成,即由一个或多个名词加上其前置定语或后置定语构成,因此标题中使用最多的是名词,其次是介词、形容词等,若出现动词,一般是现在分词、过去分词或动名词形式。

撰写标题时一个简单有效的方法是寻找出一个可以反映论文核心内容的主题词,依此进行扩展成为名词短语,使之包含论文的关键信息,并注意词语间的修饰要恰当。

(3) 标题主要有三种结构:名词词组、介词词组、名词词组+介词词组,一般不使用不定式短语或从句。

① 名词词组:由名词及其修饰语构成。

例:Severe Weather and the Automobile

　　Soil Behavior and Soil Mechanics(土壤力学)

② 介词短语(使用较少),一般使用的介词为 on,表示对……的研究。

例:On the Distribution of Sound in a Corridor

　　On the Fatigue Life Prediction of Spot Welded Components

③ 名词词组(名词)+介词词组,介词短语一般用来修饰名词或名词词组,从而限定某研究课题的范围。

例:Progress on Fuel Cell and Its Materials

　　An Analysis of the Modes of Operation of a Simple Transistor Oscillator

　　A Discussion on Self-adaptive System

　　The Application of the Models of Nonlinear Regression

标题的书写形式有三种,其中前两种使用较广泛:

① 除第一个字母、专有名词大写外,其余均小写,要求这种格式的刊物有 *Remote Sensing*,《西电学报》等。

② 开头的字母、实词及多于 5 个字母的介词、连词的第一个字母大写,其余小写,要求这种格式的刊物有《西工大学报》《西交大学报》《电子学报》等。

③ 全部字母均大写。

讨论下列标题是否合适,如何修改:
Electromagnetic Fields Have Harmful Effects on Humans
How to Use Water Resources for Irrigation in Semiarid Land
Diamond Is Used for Electronic Devices
Water Quality Can Be Protected Through the Successful Integration of Research and Education

A.3.2　作者和联系方式

论文的署名表明作者享有著作权,获得相应的荣誉和利益,但也表示文责自负,即所有作者均有义务对所发表的研究成果的科学性和真实性负责。论文署名还便于作者与同行或读者进行研讨和联系,因此有必要提供作者尽可能详细的情况,如工作单位和通讯地址,包括邮政编码等,例如:

Seismic Data De-Noising Based on Wavelet Transform

Fu Yan[1,2], Zhao Rongchun[1]

1. Dept. of Computer, Northwestern Polytechnic University, Xi'an 710072, China
2. Dept of Computer, Xi' University of Science & Technology, Xi' 710054, China

A.3.3　摘要

摘要一般在论文完成后撰写,因此我们在全文各部分介绍完之后,再介绍摘要的写法。

A.3.4　关键词

为便于读者选读及检索文献,每篇论文在摘要后应给出 3~8 个英文关键词(中文一般是 3~5 个),其作用是反映主题。

关键词的选取要注意使读者能据此大致判断论文研究内容,并且一般按照研究目的—研究方法—研究结果的顺序标引关键词。

中文科技论文应该特别注意中、英文关键词要一致:既要使表达的意思一致,又要使排列顺序一致。

A.3.5　引言

引言的主要任务是向读者勾勒出全文的基本内容和轮廓。

A.3.5.1　引言的内容和结构布局

它可以包括以下 5 项内容中的全部或几项:
(1) 介绍某研究领域的背景、意义、发展状况、目前水平等;
(2) 对相关领域的文献进行回顾和综述,包括前人的研究成果、已经解决的问题，并适当加以评价或比较；

（3）提出前人尚未解决的问题，也可提出新问题，解决这些新问题的新方法、新思路，从而引出自己研究课题的动机和意义；

（4）说明自己研究课题的目的；

（5）概括论文的主要内容，或勾勒其大体轮廓。

可以将引言分为3~4个层次来安排：

例：① Introducing the general research area including its background, importance and present level of development ... Reviewing previous research in this areas ... ② Indicating the problem that has not been solved by previous research, raising are levant question ... ③ Specifying the purpose of your research ... ④ Announcing your major findings ... Outline the contents of your paper ...

引言中各层次所占篇幅可以有很大差别，其中第一个层次会占去大部分篇幅，研究背景和目前研究状况会比较详细，研究目的会比较简短。

比较简短的论文，引言也可以相对比较简短：

例：① Introducing the importance of the research area and reviewing previous research ... ② Indicating the problem that has not been solved by previous research, raising a relevant question ... ③ Specifying the purpose of your research ...

引言的各个层次不仅有其各自的任务和目的，在语言上也有各自的特点，下面分别介绍。

A.3.5.2　引言开头的写法

引言的最主要目的是告诉读者论文所涉及的研究领域及意义是什么，即要回答以下问题：

What is the subject of the research?

What is the importance of this subject?

How is the research going at present?

In what way is it important, interesting and worthy studying?

What problem does the research solve?

因此，在写引言开头时，关键词往往出现在第一句话，迅速将主题告诉读者，然后简单介绍该研究领域的意义。

例：<u>Forecast of the tracks of hurricanes</u> have improved steadily over the past three decades. These improvements have brought about a significant decline in loss of life.

其次，引言开头句子的谓语动词一般是一般现在时或现在完成时，有一些常用

的句型,例如:

例1:研究主题+be

Spot welding is the most widely used joining method in the automobile manufacturing industry.

例2:研究主题+has become

Forest decline has become a favorite topic for environmental studies.

例3:研究主题+被动语态

Air pollution has been extensively studied in recent years.

例4:Recently, there has been growing interest in/concern about+研究主题

In the 1990s, there has been growing interest in the development of electric vehicles in response to the public demand for cleaner air.

例5:Researchers have recently focused their attention on+研究领域

或 Researchers are recently paying attention to+研究领域

A.3.5.3 引言文献综述的写法

文献综述是作者对他人在某研究领域所做的工作和研究成果的总结和评述,包括他人有代表性的观点或理论、发明及发现、解决问题的方法等。

在援引他人的研究成果时,必须标注出处。标注出处时有两种方法:一种是著者-出版年,另一种是顺序编码。

例1:Hanson et al. (1976) presented this new method.

This new method was presented by Hanson et al. (1976).

例2:A new method was presented by Hanson et al.[1](或 et al.[1—3])

如果需要多处引用同一个作者或同一篇文章,那么需要使用一些连接手段使上下文衔接。

例:The author goes on to say that ...

The article further states that ...

The author also argues that ...

A.3.5.4 研究动机与研究目的的写法

介绍研究动机可从两个角度入手:一是指出前人尚未解决的问题,二是说明解决这一问题的重要意义。在指出前人尚未解决的问题时,有一些常用句型,下面进行较详细的介绍。

用表示否定意义的词(如 little, few, no 或 none of)+名词作主语:

Little information/work/research ...

Few studies/investigations/researchers ...

No studies/data ...

None of these studies/findings …

例：There have been few specific reports in the literature of oak and hickory decline.

用表示对照的句型：

The research has tended to focus on … , rather than on …

Although considerable research has been devoted to … , rather less attention has been paid to …

提出问题或假设：

However, it remains unclear whether …

If these results could be confirmed, they would provide strong evidence for …

提出问题后下一步应指出本研究的目的和内容，可使用以下句型：

描述研究目的最简单的句型是用 this paper、this study、this project、this research 等词作主语，后面用 investigate、discuss、examine 等动词作谓语，宾语为研究内容，即：

This paper
This study
This project
The present study
This research
Our project
This survey
This thesis

{ concerns / tests / investigates / reports / discusses / describes / explains / calculates / examines / analyses / proposes / demonstrates / measures } …

把论文本身当作强调的内容，通常采用一般现在时：

The purpose of this paper is to discuss …

The aim of this report is to evaluate …

The objective of the present paper is to examine …

把研究活动作为主语，通常采用被动语态：

This research is designed to determine …

This study is designed to measure ...

Our project aim to calculate ...

A.3.5.5 引言结尾的写法

写引言结尾时可以把研究目的作为引言的结尾,也可以简单介绍文章的结构和各部分的主要内容,使读者了解文章的轮廓。

介绍文章结构时注意避免使用同一个句型结构,如 Section 1 describes ... , Section 2 analyses ... , Section 3 discusses ... 等。

可以使用如下结构进行替换:

... Section 2 defines ... , ... is then presented in section 3, followed by ... , ... is evaluated in section 4, while section 5 discusses ... . Finally, in section 6, ... is given.

A.3.6 论文主体

科技论文按内容可分为实验型、理论型和综述型,下面分别介绍。

A.3.6.1 实验型论文的论文主体

实验型论文阐述的核心内容是实验,以及进一步对实验结果进行定性或定量的讨论。实验型论文的主要内容如下:

(1) 实验用原料和材料

撰写实验用原料和材料的基本原则为清晰、准确与简洁,为读者能再现论文中的研究结果提供必要条件。通常是先概述所用的原料和材料,再详细描述其结构、成分、特性与功能等。

(2) 实验技术和方法

撰写实验技术部分时要列出实验所用的主要的、关键的实验设备和仪器及操作过程,并说明研究过程中实验条件的变化因素及其考虑的依据和设想等。实验部分还要注意实验的设计,用最少的实验次数获得最优的实验参数组合。

撰写实验方法部分时总体要把握重点突出、详略得当:对公知、公用的方法写明其方法名称即可;对已有应用但尚未为人们所熟悉的新方法应注明文献出处,并对其方法做简要介绍;对改进或创新部分应详细介绍。

总之,实验技术和方法的介绍,既便于为其他研究者提供一个可重复研究的依据,又可以提高读者对该研究设计及其结果可靠性的信任程度。

(3) 实验结果

实验结果是对研究中所发现的重要现象的归纳,论文的讨论由此引发,对问题的判断推理由此导出,全文的一切结论由此得到。撰写实验结果部分时重点要向读者提供对结果的介绍、描述及评论等信息,使读者在论文前面部分(引言、材料和方法)基础上,对获得的结果有一定的客观认识。

（4）讨论

讨论是论文的重要部分,对读者具有启迪作用,也是比较难写的部分。在这部分中,作者要回答引言中所提的问题,评估研究结果所蕴涵的意义,用结果去论证问题的答案。讨论部分写得好会充分体现论文的价值。

撰写讨论部分的步骤通常是:

再次概述研究目的或假设,说明预期结果是否实现,提出问题,然后简述最重要的结果,并指出这些结果能否支持原先的假设或是否去其他研究者的结果一致,有时还会再次强调个别的重要结果。

对结果进行分析、说明、比较和评价,总结与他人的研究相比本论文研究的特色,给出由结果所能得出的推论或结论并指出研究方法或结果的局限性及由此对结果所产生的影响。

点出研究结果的理论意义或实际应用价值。

A.3.6.2 理论型论文的论文主体

理论型论文是以理论研究为主体的学术论文,在对它进行理论阐述时要注意逻辑推理。理论阐述包括论证的理论依据,对其所做的假设与合理性进行的论证,对于分析方法所做的说明。在写作时,应注意区别哪些是已知的,哪些是作者第一次提出的,哪些是经过作者改进的,等等,这些都应详细说明。

总之,理论阐述的要点是:假设,前提条件,分析的对象,所引用的数据及其可靠性,适用的理论或新模型的提出,分析方法,计算过程,新理论或模型的验证,导出的结论。

A.3.6.3 综述型论文的论文主体

综述型论文的内容一般包括:① 问题的提出;② 发展历史的介绍;③ 现状分析;④ 未来发展趋势和建议。

综述型论文的价值在于,作者通过对已有文献的综合和归纳,分析某学术领域或某一方面研究工作的发展历史和现状,指出该领域科学活动的发展方向,提出具有科学性、创造性和前瞻性的研究课题。

下面是论文主体中的一些常用句型:

（1）描述实验方法常用的句型

描述实验方法与过程的句型通常采用被动语态的过去式,因为在描述实验过程或方法时,句子中的主题或中心是实验材料、场地和方法本身,表达"做了什么""怎么做的"之意,而不是表达"谁做了什么"。因此,在描述方法时,常将实验材料等作为主语,谓语动词自然要用被动语态。例如:

... the sections were rinsed to remove adherent material and dried ...

The adventitious roots(不定根)from each section were removed ...

Five agents were selected at random and asked to collect a water sample for the instructor.

（2）理论分析的常用句型

在正文部分的假设、理论分析、数理模型的建立以及计算过程中，常常要推导并描述公式，作者应了解和掌握公式的规范表达和描述公式的基本句型。科技论文理论分析部分中资料大部分表达数字与逻辑关系，这类表述是不受时间影响的普遍事实，所以表示理论分析的常用句型使用一般现在时态。例如：

Substituting M in/into N, we obtain/have/get ...

Substituting M in/into N gives/yields/results in ...

Let us now consider the case the dynamic link parameters are unknown.

Suppose that X_t is a solution of the SDE(1).

If $m=1.2$, then we have the following equations ...

Given that $m=1.2$, we obtain ...

The relationship between m and n is as follows ...

描述公式时需注意以下几点：

每个公式占一行，居中；

有两个或两个以上公式时，要有编号，编号加上圆括号，列在右端；

行文中，将公式看作一个单词处理，若是一句话的结束，需加句点，否则根据情况选用合适的标点；

若公式中符号是第一次出现的，要加以解释，用 where 或 in which 引导。公式中符号也作为一个单词处理，所以符号后谓语动词用单数形式。

行文中涉及某一公式时，要用 Equation 或 Equ 表示，例如：

The current in the wire is calculated using

$$V=IR,$$

where(in which) V is voltage, I is current, and R is resistance.

By eliminating l between Equations(1) and (2), we have ...

Let f be a continuous function, ...

Suppose that g is a constant, ...

When x is close to 0, then y is close to 1.

（3）图表文字说明中的常用句型

为了清楚地描述研究结果，作者可能将结果用图表表示。注意表头要写在表上面，图题要写在图下面。图表中的内容要用文字来说明。在文字说明中，如果提及图表，表格用"Table＋序号"表示，如"Table 1 shows ... "。其他各种类型的图可以用"Figure＋序号"表示，如"From Figure 2 we learn ... "。"Figure"这个词可以用缩写"Fig."，但必须前后统一。

文字说明中的动词要用一般现在时态，选用的动词要慎重，因为不同的动词表示研究结果的可靠性不同，同时也表示作者不同的判断、推理和态度。例如：

Table 1 shows/provides details of the test results.

Figure 4.2 gives the results of the experiment.

Both of these questions received very high ratings (see Table 1).

上述句型也可使用被动语态，例如：

As shown in Table 3, ...

As can be seen from the data in Table 1, ...

As described in Figure 3.2, ...

（4）介绍研究成果的常用句型

作者对其研究成果的评论或说明通常包含下列内容之一。

根据本人的研究结果作出推论。如：

These results suggest that untrained octopuses（章鱼）can learn a task more quickly by observing the behavior of another octopus than by reward-and-punishment methods.

作者解释研究结果或说明产生研究结果的原因。如：

These findings are understandable because the initial annealing（退火）temperature dictates（规定）the state of conformational（构象的）structures.

作者对此次研究结果与其他研究者曾发生的结果作比较，例如指出自己的结果是否与其他研究者的结果一致：

These results agree with Gerner's analysis, in that Q_{max} varies inversely with length and to the third power of the pipe width.

作者对自己的研究方法或技术性能与其他研究者的方法或技术性能进行比较。如：

The recognition rate of our system is significantly higher than that reported for Token's system.

作者指出自己的理论模型是否与实验数据符合。例如：

The measured temperatures along the heat pipe are all highly consistent with the predictions of the theoretical model.

当评论的内容为对研究结果可能性的证明时，句子的主要动词之前通常加上 may 或 can 等一般现在时态的情态动词。例如：

The layer structural or some other mixed complex material may be the most suitable refrigerants for the Ericsson magnetic refrigerator.

One reason of this advantage may be that the Hank visual programming language can avoid some of the syntactic problems associated with textual programming languages.

A possible explanation for this is that …

These results agree well with the findings of Cassdio, et al.

当作者指出由自己的理论模型所得到的预测与实验数据之间的吻合程度时，通常使用一般现在时态，说明模型预测与实验数据是否一致不受时间影响。例如：

The data confirm closely the prediction of the model.

The theoretical model fits the experimental data well.

The theoretical model agrees well with the experimental data.

The experimental measurements are very close to the predicted values.

（5）讨论部分的常用句型

在讨论部分中，作者通常对研究结果进行概述、分析、解释等，因为陈述的是作者的见解和结论，所以句型多用一般现在时态。

概述结果：

These results provide substantial evidence for the original assumptions.

These experimental results support the original hypothesis that …

These results contradict the original hypothesis.

The present results are consistent with those reported in our earlier work.

表示研究的局限性：

It should be noted that this study has examined only …

The findings of this study are restricted to …

The limitations of this study are clear …

The result of the study cannot be taken as evidence for …

Unfortunately, we are unable to determine from this data …

A.3.7 结论

结论是作者对有关研究课题进行的总体性讨论，必须具有严密的科学性和客观性，反映课题研究价值，并对以后的研究具有指导意义。

A.3.7.1 结论部分的内容与结构布局

结论部分一般包括以下 5 项内容：

（1）概括说明课题的研究内容、结果及其意义与价值；

（2）较具体说明课题证明了什么假设或理论，得出什么结果，有何创造性成果或见解，解决什么问题等；

（3）与他人的相关研究进行比较；

(4) 课题的局限性、不足之处、有待解决的问题等；

(5) 指出进一步研究的方向。

上述五项中，前两项必不可少，后三项可根据需要进行取舍。

下面给出一个例子：

Conclusions

Methods for speckle reduction and enhancement of SAR images using DMWT have been proposed. The new methods significantly reduce the speckle while preserving the resolution and the structure of the original SAR images. Cleaner images should improve classification and recognition.

A.3.7.2 结论部分的语言运用技巧

结论部分的语言除要连贯、精练、确切以外，还要具有高度的概括性。结论部分中，现在时态使用频率很高，这是因为结论部分总结研究者到目前为止做了哪些工作，得出了什么结果，这些结果对现在来说有什么影响、意义、价值，能够用于什么地方，解决什么问题等。例如：

Two factors to influence mold filling have been studied …

Through the example of a 60-storey building, it has demonstrated that a simplified approach can be used …

Overall, our study has revealed a variety of patterns at the community and population levels.

作者对结论部分进行说明时，有时对研究结果所持态度明确、肯定，有时不那么肯定，这时往往对动词加以修饰，表明作者不太肯定的态度，或缓和语气。

(1) 表示对结果肯定的情况

This study clearly demonstrates that …

The investigation discovers that …

This study reveals that …

These results represent …

(2) 表示对结果不太肯定的情况

表示对结果持不太肯定的态度，可以使用情态动词、某些形容词或副词，或用 appear、seem 等动词表示。

① 用情态动词

最常用的情态动词有 will、can、may 及其过去式 would、could、might 等，三个情态动词语气依次减弱，过去式比现在式语气弱。例如：

Changes occurring due to insects and pathogens may be a natural and necessary phenomenon.

By utilizing these methods, more teachers may be able to achieve their aims.

Evaluation of decline symptoms in smaller size classes might provide additional useful information to be used in understanding the oak regeneration problem.

② 用某些形容词或副词

不同的形容词和副词表达作者不同的态度和语气，如 certainly、of course、absolutely、definitely、clearly 等用来加强语气，表示作者态度肯定；almost、probably 次之，而 likely、unlikely、perhaps 等词表示作者没有把握，语气比较弱。例如：

Evaluation of decline symptoms in smaller size classes are likely to provide additional useful information to be used in understanding the oak regeneration problem.

Perhaps changes occurring due to insects and pathogens are a natural and necessary phenomenon.

③ 用 seem、appear、tend 等动词

这三个动词用在另一个动词之前，构成 seem to＋do/be/have，appear(tend) to ＋ do/be/have 的结构，表示陈述的现象、内容或观点有进一步探讨的余地。例如：

Science projects appear to be an important part of science education.

下面句子的语气依次减弱：

Science projects are an important part of science education.

Science projects tend to be an important part of science education.

Science projects appear to be an important part of science education.

Science projects seem to be an important part of science education.

It would appear that science projects are an important part of science education.

在分析、综合、归纳、推理过程中，常会使用 show、prove、illustrate、reveal、suggest、indicate、clarity、verify 等动词，这些词含义不同，表示的语气和态度也不同，注意区分使用。例如：

Clearly, sporadic(零星的)insect outbreaks, pathogens and climatic events have caused excessive mortality(死亡率) of certain species.（表示直接原因，负全部责任）

Clearly, sporadic insect outbreaks, pathogens and climatic events have lead to excessive mortality of certain species.（表示间接原因）

Clearly, sporadic insect outbreaks, pathogens and climatic events have contributed to excessive mortality of certain species. （表示部分原因，还有别的原因）

A.3.8　附录

对于实验测得的重要原始数据、有代表性的计算实例、重要的公式推导和主要设备的技术性能、程序代码等，有时编入正文会有损逻辑性，但这些又是与正文有密切联系且具重要参考价值的材料，此时可将它们放入附录，用以帮助读者更好地掌握和理解正文内容。一般科技期刊不提倡另列附录，只有在学位论文中，由于没有字数限制，附录较为常见。

A.3.9　致谢

致谢对象有两类：一类是经费上给予支持的，另一类是在技术、方法、条件、资料、信息等方面给予建议和帮助的。致谢时一般项目资助单位或个人在前，其他人员在后。

致谢中常用句型如下：

（1）感谢资助单位或个人

Support for this program/project/study is provided by …

Funding for this program is provided by …

I thank … for giving financial support for this study.

Research for this paper was partially supported by …

This research was funded by …

（2）感谢提供资料的单位或个人

Data were supplied/provided by …

Permission to quote from material protected by copyright has been granted by …

I thank … for the permission to quote from material protected by copyright.

（3）感谢给予建议和帮助的个人

I thank … for comments on the manuscript.

In addition, I am grateful to … for their valuable suggestions, and to … for her patience and good counsel.

For help in the technical assistance, I thank …

We thank the following people for their helpful comments on drafts of this paper：…

In addition, I wish to thank … for his valuable suggestions.

For their encouragement, support, and research assistance, I would like to thank the following individuals who have contributed substantially to the completion of this study/paper/work.

(4) 感谢不知名的审稿人

The anonymous reviewers have also contributed considerably to the publication of this paper.

In addition, I would like to thank the anonymous reviewers who have helped to improve the paper.

(5) 感谢其他人

I also owe an obligation to …

Also I wish to thank … for their many courtesies.

Thanks is also extended to … who …

I wish to thank … for his …

A.3.10　参考文献

在论文的最后列出参考文献,其目的在于:① 便于读者查阅原始资料中的有关内容,以了解前人工作与作者工作的区别及本论文的起点;② 便于读者了解论文工作的基础,并了解其理论基础优劣;③ 尊重他人的劳动成果,突出论文的创新性;④ 将论文成果和观点与前人的研究作比较,同时也说明论文的创新性;⑤ 有利于缩短论文的篇幅。

需要注意,文后列出的参考文献应是公开正式发表过的、作者真正参阅过的、与论文密切相关或直接引用的文献。另外,引用文献时要注意引用最新的文献,它标志着论文作者对最新科研工作的了解及掌握情况,避免重复性工作出现,体现论文的创新性。

A.4　Abstract Writing　摘要的撰写

摘要作为对研究论文正文的精练概括,便于读者在最短时间内了解全文内容。摘要可以独立于正文,通常收录于相应学科的摘要检索类数据库或专刊内,撰写好摘要对于论文是否被数据库收录和他人引用至关重要。

摘要的目的是为读者提供关于文献内容的有用信息,即论文所包含的主要概念和所讨论的主要问题。读者从摘要中可获知作者的主要研究活动、研究方法和主要研究结果及结论。摘要可以帮助读者判断此论文对自己的研究工作是否有用,是否有必要获取全文,为科研人员、科技情报人员及计算机检索提供方便。

A.4.1 摘要的种类和特点

摘要主要有以下三种：

(1) 随同科技论文一起在学术刊物上发表的摘要。置于论文的主体部分之前，目的是让读者首先了解一下论文的内容，以便决定是否阅读全文。一般这种摘要在全文完成之后写，一般在 100～150 个单词范围，但也有字数很长的摘要。本类摘要内容包括研究目的、研究方法、研究结果和主要讨论。

(2) 学术会议论文摘要。往往在会议召开几个月前撰写，交由会议论文评审委员会评阅，从而决定能否录用。此类摘要比第一种摘要略为详细，长度在 200～300 个单词范围。开头有必要介绍一下研究课题的意义、目的、宗旨等。如果在写摘要时研究工作尚未完成，全部研究成果还未得到，那么应在方法、目的、宗旨、假设方面多花笔墨。

(3) 学位论文摘要。一般在 400 个单词左右，可根据需要分为几个段落。内容一般包括研究背景、意义、主旨和目的；基本理论依据、基本假设；研究方法；研究结果；主要创新点；简短讨论。要突出创新之处，指出有何新的观点、见解或解决问题的新方法。

以上四种摘要具有共性，内容一般都包括：① 目的（objectives or purposes）：包括研究背景、范围、内容、要解决的问题及解决这一问题的重要性和意义；② 方法和材料（methods and materials）：包括材料、手段和过程；③ 结果与讨论（results and discussions）：包括数据和分析；④ 结论（conclusions）：主要结论、研究的价值和意义等。

摘要的撰写要求：

(1) 摘要应具有独立性和自明性，并拥有文献的主要信息，即不阅读文献的全文，就能获得必要的信息。因此，摘要是一种可以被引用的完整短文。

(2) 用第三人称。作为一种可阅读和检索的独立使用的文体，摘要主要用第三人称而不用其他人称来写。有的摘要使用"我们""作者"作为陈述的主语，一般来说，这会减弱摘要表述的客观性，有时也会出现逻辑上不通的问题。

(3) 排除在本学科领域方面已成为常识的内容。

(4) 不得简单地重复论文篇名中已经表述过的信息。

(5) 要客观如实地反映原文的内容，要着重反映论文的新内容和作者特别强调的观点。

(6) 要求结构严谨、语义确切、表述简明，一般不分段落；切忌发空洞的评语，不作模棱两可的结论。

(7) 要采用规范化的名词术语。

(8) 不使用图、表或化学结构式，或相邻专业的读者尚难于清楚理解的缩略语、简称、代号。如果确有必要,在摘要中首次出现时必须加以说明。

(9) 不得使用文献中列出的章节号、图、表号、公式号以及参考文献号。

(10) 要求使用法定计量单位以及规范字和标点符号。

(11) 众所周知的国家、机构、专用术语尽可能用简称或缩写。

(12) 不进行自我评价。

A.4.2 摘要的内容和结构

(1) 学术期刊论文摘要:简短精练是其主要特点,只需简明扼要地将研究目的、方法、结果和讨论分别用 1~2 句话加以概括即可。至于研究背景和宗旨应在论文的 Introduction 中较详细介绍。这种摘要结构布局如下:

Title

Author(s), address

Objectives, purposes, hypothesis:…

Methods, materials, procedures:…

Result, data, observations, discussion:…

Conclusion:…

(2) 学术会议论文摘要:它直接决定论文是否被录用,因此首先要简要说明研究背景、内容、范围、价值和意义,在研究方法上也可以多花一些笔墨。这种摘要结构布局如下:

Title

Author(s), address

Background, previous studies, present situation, problems that need to be solved:…

Objectives of this study, hypothesis:…

Methods, materials, procedures:…

Results, data, observations, discussion:…

Conclusions:…

(3) 学位论文摘要:一般单独占一页,也可能占两页,装订在学位论文目录之前。学位论文摘要也要介绍研究背景、内容、目的、方法、结果等。但是学位论文摘要的不同之处在于它必须指出研究结果的独到之处或创新点。关于研究的内容也可稍加详细介绍。这种摘要结构布局如下:

Title

Author(s), address

Background, problems that need to be solved, rationale for the present study：...

　　Objectives and scope of this study：...

　　Outline of the main contents and results：...

　　Conclusion：...

下面给出一个期刊论文摘要的例子：

Does Aluminum Enter the Liquid Contained in Pop Cans and Aluminum Cookware

Daniel T. Moss

... （address）

　　This investigation was performed to see if the aluminum in pop cans and aluminum cookware enters the liquid they contain. It was hypothesized that aluminum does enter the liquids in aluminum cans and cookware. This experiment was performed by testing seven different types of carbonated beverages and water boiled in aluminum cookware for three hours. Every hour a sample was removed for testing. The cookware consisted of some new and old aluminum pots. The colorimetric method was used to determine the concentration of aluminum. The results showed that aluminum was present in carbonated beverages and in the water boiled in the cookware. It was concluded that aluminum existed in carbonated beverages and substances cooked in aluminum cookware, and that the concentration of aluminum increased with time.

　　此前，西北工业大学的第一作者论文被 IAA（International Academy of Astronautics，国际宇宙航行学会）收录的最高摘要行数纪录是 23 行，这个纪录到 2001 年底都没有被打破，因为破纪录的难度很大。而在 2002 年 IAA 收录了 2 篇平校纪录的西工大第一作者论文，两篇论文的作者分别为李杰和杜随更。其中，李杰在 2003 年被 IAA 收录的论文摘要行数达到 25 行，破了校纪录。下面是胡沛泉老先生对李杰的英文摘要的评价：

　　"李杰能写很好的英文摘要，他所写的英文摘要只用修改几分钟就好了。西工大作者能写很好的英文摘要的很少，李杰的英文摘要质量之高在西工大是突出的。李杰个人的努力，再加上他在西工大的本科及研究生阶段所处的环境特别有利于他提高英语水平，另外他的研究成果是在两位教授多年的研究基础上进行的，因此论文内容在国际上具有较强的竞争力。"

　　杜随更个人写作英文摘要的能力较弱，但他花了颇多时间来努力写好英文摘要并且向胡先生详细说明了他的论文的实际意义。他的做法是不依赖胡先生，在自力更生方面做最大努力，但也积极向胡先生请教。

杜随更英文摘要修改随感：

"修改时，胡先生先重写了英文题目，将原题目 *Three-dimensional FEM Analysis Coupled with Thermo-mechanical During Linear Friction Welding*（线性摩擦焊接过程三维热力耦合有限元分析）修改为 *On Exploring Linear Friction Welding of Blade to Disk of Aeroengine Rotor*（有关航空发动机整体叶盘的线性摩擦焊接过程的研究），强调了研究背景，增加了读者的兴趣度与读者群的范围。

胡先生从以下几个方面对摘要进行了修改：

（1）研究背景：摘要开首即点明"Blisk（integrally bladed disk）is a key technology for improving performance of aeroengine"，并指出"When the blade material is … there are unanswered questions that need to be explored about linear friction welding of blisk"，说明了研究的必要性。

（2）研究方法：用三个词"Theoretical exploration requires"引出了本文的研究方法，特别强调了"Different from usual theoretical simulation, we simulate directly from the parameters of the welded materials and the welding conditions"，简明扼要地指出了该文的研究特点。

（3）研究结果：在简介了研究背景及方法之后，大篇幅地介绍了研究结果和结论，反映出了该文的研究精髓。"

李杰博士写好文章的英文摘要的几点经验：

（1）长期、大量地阅读相关的专业英文文献是写好英文摘要的基础工作，也是必不可少的工作。只有通过大量的阅读，才能熟悉英文科技文献的写作习惯，掌握专业术语的习惯用法。

（2）平时阅读外文资料时，注意对一些专业术语、叙述技巧的积累。日积月累，对一些问题的叙述，自然比较贴近国外的习惯。

（3）在写英文摘要时，不要根据已有的中文稿件逐句、逐段翻译。英文文章亦是如此。因为中文和英文的写作习惯、侧重点不同，如果只是按字面意思翻译，难免会限制作者的思路，使译出来的东西显得生硬、晦涩。

（4）不要等到一项科研课题完全结束时，才对所取得的成果进行总结。在研究取得阶段性进展时，就可对一些内容进行总结，随着研究工作的深入，再对文章进行充实、修改。

（5）要写好文章，须请英文水平比较高的老师、同事指正，帮助发现问题，同时，也要找英语水平一般的学生阅读，多听取他们的意见。发表文章的目的在于交流，如果某种叙述，大多数人看不明白，就说明自己在表达方面肯定存在一定的问题。

胡沛泉老先生有关英文摘要修改方面的经验：

（1）不能因为写不好英文摘要，就有意把它写得太短，这样会造成文章创新点含糊不清，论文重要性无法得到体现。

（2）写摘要时要在研究目的、研究方法、研究结果三方面重点书写，内容要丰富，体现出研究的价值。

（3）在写摘要之前先清楚论文所涉及课题的背景意义、论文的主要观点结论及论文的重要性，总结出读者对该研究可能感兴趣的几个方面。

胡老先生在2003年12月18日修改闫守孟撰写的英文摘要时发现其写得较好，能体现论文价值，但原来的英文题目太平淡，和作者讨论后换了一个题目，并在摘要开头加了一句话：面向嵌入式驱动软件开发的设备仿真技术研究。该摘要节选如下：

摘要：分析了传统嵌入式驱动软件开发模式的不足，提出了一种基于仿真设备的软硬件协同开发模式，详细论述了设备仿真技术的具体设计与实现，结合应用实例证明了该技术能较有效地缩短嵌入式驱动软件开发周期。

A New Device Simulation Method for Shortening Development Period of Embedded System

Abstract：Existing device simulation methods for embedded driver development, to our best knowledge, do not pay enough attention to computer hardware simulation, thus making development period of embedded system unnecessarily long ...

再看一个例子：

小波变换在地震信号噪声处理中的应用

摘要：常规小波域阈值去噪方法未能充分利用地震信号相关性的特点进行去噪，只能去除地震信号中部分随机噪声，为此本文提出一种小波域分时分频相关结合阈值去噪处理方法。该方法首先对小波变换后多个尺度上小波系数进行分时分频相关去噪处理，然后对处理后小波系数进行重构，该方法可去除大部分不相关随机噪声。对重构后地震信号再进行常规小波域阈值去噪处理以进一步去除噪声。模型测试和实际资料处理效果表明：使用本文方法可以有效改进地震信号去噪处理效果。

Seismic Data De-noising Based on Wavelet Transform

Abstract：Conventional de-noising method by threshold filter in wavelet domain cannot utilize the correlations of seismic data to remove random noises, so a new method which is named de-noising method by combine time-frequency correlation analysis with threshold filter in wavelet domain is proposed. By using this method, time-frequency correlation analysis was made on wavelet coefficients of multi-scales firstly, then construct these wavelet coefficients, most random noises can be removed. Secondly using conventional de-noising method by threshold filter in wavelet domain to remove more noises. The results show that the new method can improve the effect of disposal and improve signal-to-noise ratio by processing the synthetic data and real data.

修改后的摘要：

Abstract：Conventional de-noising method by threshold filter in wavelet domain do not utilize the correlations of seismic data to remove random noises. So we propose a new and we believe better method which we call de-noising method. This de-noising method combines time-frequency analysis with threshold filter in wavelet domain. We explain in much detail how to use time-frequency correlation analysis to analyze correlations of seismic data, i. e., to analyze wavelet coefficients of multi-scales; after correlation analysis, we reconstruct these wavelet coefficients; in this way, most random noises can be removed. We then use conventional de-noising method by threshold filter in wavelet domain to remove more noises. We applied our new de-noising method to seismic data and did succeed in removing most random noises and raising SNR (signal-to-noise ratio).

Appendix B　Academic Information for Mining

采矿专业学术信息

一、National Websites for Mining Engineering　国内采矿业网站

中国煤炭资源网：http://www.sxcoal.com
中国煤矿安全网：http://www.mkaq.org
中国煤炭信息网：http://www.coalcn.com
国家煤矿安全监察局：http://www.chinacoal-safety.gov.cn
环境健康安全网：http://www.ehs.cn/portal.php
选矿选煤网：http://www.xkxm.com
粉煤灰综合利用网：http://www.flyingash.com
中国煤炭新闻网：http://www.cwestc.com
中国煤炭工业展览网：http://www.coal-china.org.cn

二、International Websites for Mining Engineering　国际采矿业网站

美国安全工程师协会(American Society of Safety Engineers)：www.asse.org
美国矿山安全健康监察局(National Mine Safety-Health Administration)：www.msha.gov
美国能源部(USA Department of Energy)：www.energy.gov
美国职业病控制中心［USA Centers for Disease Control and Prevention，The National Institute for Occupational Safety and Health (NIOSH)］：www.cdc.gov/niosh/mining
澳大利亚联邦科学院［Commonwealth Scientific and Industrial Research Organization (CSIRO)］：www.csiro.au
采矿技术：www.mining-technology.com
岩石科学：www.rockscience.com
煤矿采矿协会：(National Mining Association)：www.nma.org
英国煤炭公司：www.ukcoal.com

维基百科：en. wikipedia. org/wiki/coal mining
美国 JOY 公司：www. joy. com
美国 CONSOL 能源公司：www. consolenergy. com
德国 DBT 公司：www. energy-mining. german-pavilion. com
煤炭教育网：www. coaleducation. org
绿色煤层气网：www. greengas. net
世界煤炭：www. worldcoal. org

三、Universities for Mining Engineering　采矿工程专业大学

1. 美国

California Polytechnic State University-San Luis Obispo

Carnegie Mellon University Civil & Environmental Engineering

Colorado School of Mines

Henry Krumb School of Mines at Columbia University

Michigan Technological University Metallurgical Engineering and Mining Engineering Dept

Montana Tech of the University of Montana

Pennsylvannia State University's College of Earth and Mineral Sciences

South Dakota School of Mines and Technology

U. C. Berkeley：Department of Materials Science and Mineral Engineering

University of Alaska Fairbanks School of Mineral Engineering

University of Arizona College of Engineering and Mines

University of Kentucky Mining Engineering Dept

University of Minnesota Newton Horace Winchell School of Earth Sciences

University of Missouri-Rolla Mining Engineering Dept

University of Nevada Reno Mackay School of Mines

University of North Dakota School of Engineering and Mines

University of Oklahoma College of Engineering

University of Utah College of Mines and Earth Sciences

University of Wisconsin-Madison

University of Wisconsin-Platteville

University of Wyoming：Department of Geology and Geophysics

Virginia Polytechnic Institute：Department of Mining and Minerals Engineering

West Virginia University College of Engineering and Mineral Resources

2. 加拿大

Laurentian University School of Engineering

Laval University: Department of Mining and Metallurgy

McGill University: Department of Mining and Metallurgy

Queens University Metallurgy Dept.

Queens University Mining Dept.

Technical University of Nova Scotia: Department of Mining and Metallurgical Engineering University of Alberta Faculty of Engineering

University of British Columbia Faculty of Applied Science

University of Calgary: Department of Geology and Geophysics

3. 澳大利亚

Australian National University: Department of Geology

Central Queensland Institute of Tafe

Flinders University Geoscience Switchboard

James Cook University: Department of Earth Sciences

La Trobe University School of Earth Sciences

Macquarie University School of Earth Sciences

Monash University: Department of Earth Sciences

Queensland University of Technology: Department of Geology

Royal Melbourne Institute of Technology

University of Ballarat University of Melbourne School of Earth Sciences

University of New England: Department of Geology and Geophysics

University of New South Wales School of Mines

University of Newcastle Geology Dept.

University of Queensland: Department of Mining and Metallurgical Engineering

University of Tasmania Geology Dept.

University of Western Australia: Department of Geology and Geophysics Western Australian School of Mines

4. 波兰

Stanislaw Staszic University of Mining and Metallurgy in Cracow

5. 俄罗斯

Russian Academy of Sciences: Department of Geology, Geophysics, Geochemistry and Mining Sciences

6. 南非

University of the Witwatersrand: Department of Mining Engineering

四、International Journals for Mining Engineering　国际采矿期刊

International Journal of Rock Mechanics and Mining Science
International Journal of Coal Geological
Journal of Rock Mechanics and Geotechnical Engineering
Mining Science and Technology
Rock Mechanics and Rock Engineering
Journal of Mining Science
Journal of Coal Science and Engineering

References 参考文献

[1] 全国自然科学名词审定委员会. 煤炭科技名词[M]. 北京:科学出版社,1996.

[2] 蒋国安. 采矿工程英语[M]. 2版. 徐州:中国矿业大学出版社,2011.

[3] 周科平,李杰林. 采矿工程专业英语[M]. 长沙:中南大学出版社,2010.

[4] 袁亮. 低透气性煤层群无煤柱煤与瓦斯共采理论与实践[M]. 北京:煤炭工业出版社,2008.

[5] Peng S S. Surface Subsidence Engineering[M]. New York:Society for Mining Metallurgy and Exploration,1992.

[6] Peng S S. Longwall Mining[M]. Morgantown:West Virginia University,2006.

[7] Peng S S. Ground Control Failure:A Pictorial View of Case Studies[M]. Morgantown:West Virginia University,2007.

[8] Peng S S. Coal Mine Ground Control[M]. Morgantown:West Virginia University,2008.

[9] Stefanko R. Coal Mining Technology Theory and Practice[M]. New York:Society of Mining Engineers of the American Institute of Mining,Metallurgical,and Petroleum Engineers,1983.

[10] Top Ten Coal Producers from World Coal Association 2018[EB/OL]. https://www.worldcoal.org/coal/coal-mining.

[11] Hoek E. Practical Rock Engineering[M]. North Vancouver:Evert Hoek Consulting Engineer Inc.,2006.

[12] Crickmer D F,Zegeer D A. Elements of Practical Coal Mining[M]. New York:Society of Mining Engineers of the American Institute of Mining,Metallurgical,and Petroleum Engineers,1981.

[13] Hustrlid W A,Bullock R L. Underground Mining Methods[M]. New York:Society for Mining Metallurgy,and Exploration,Inc.,2001.

[14] Brady B H G,Brown E T. Rock Mechanics for Underground Mining[M]. London:George Allen and Unwin,1985.

[15] Hagan T N. Controlling Blast Induced Cracking around Large Caverns[C]//Proc. ISRM symp., Rock Mechanics Related to Caverns and Pressure Shafts. Aachen, West Germany, 1982.

[16] Pokrovskii N M. Driving Horizontal Workings and Tunnels [M]. Moscow: Mir Publishers, 1980.

[17] Chase F E, Newman D, Rusnak J. Coal Mine Geology in the U. S. CoalFields: A State of Apt[C]//Proceedings: 25th International Conference on Ground Control in Mining. Morgantown: Dept. of Mining Engineering, College of Mineral and Energy Resources, West Virginia University, 1987.

[18] Fassett J E, Durrani N A. Geology and Coal Resources of the Thar CoalField, Sindh Province, Pakistan[R]. U. S. Department of Interior, 1994.

[19] Lukowicz K. Using Overlaying Methane Drainage of Active Longwalls Goaf and Its Influence on Methane Hazard Level in Coal Mines[D]. Central Mining Institute, 1998.

[20] Kay D, Waddington A, Page J, et al. Management of Impacts of Longwall Mining under Urban Areas[C] //Coal 2006: Coal Operators Conference, University of Wollongong & the Australasian Institute of Mining and Mletallurgy, 2006.

[21] Leonard II J W. Coal Preparation[M]. 5ed. New York: Society for Mining Metallurgy, 1991.

[22] Crowell D L. Coal Mining and Reclamation[Z]. Ohio: Geo Facts. Ohio Division of Geological Survey, Ohio Department of Natural Resources, 2001.

[23] Zhang H X, Zhao Y X. State of Art and Development Current of Strip Pillar Mining [J]. Coal Mining Technology, 2000: 35-37.

[24] Guo W B, Chai Y Y. The Photo Elastic Experiment Study on the Stress Distributicn Law of Strip Mining [J]. Journal of Liaoning Technical University, 1998,17(6):590-594.

[25] Sui H Q, Wang Z L. Principle and Application of Surface Subsiden Cecontrolled by Grouting in Overlying Separation Layer[J]. Chinese Journal of Geotechnical Engineering, 2001,23(4): 510-513.

[26] Qian M G, Xu J L, Miu X X. Green Technique in Coal Mining[J]. Journal of China University of Mining and Technology, 2003, 32(4): 343- 348.

[27] Bossard F C. Chap. 22: Primary Mine Ventilation Systems[C]//A Manual of Mine Ventilation Design Practices. Butte, MT: Floyd C. Bossard &. Asso., Inc.,1982.

[28] Cecala A B. Determining Face Methane Liberation Patterns During Longwall Mining[C]//Proc. 2nd U. S. Mine Vent. Sympo. Rotterdam: Mousset-Jones Press,1985: 361-367.

[29] Colinet J F, Spencer E R, Jankowski R A. Status of Dust Control Technology on USA Longwalls[C]//Proc. 6th Int'l Mine Vent. Cong. New York: Society for Mining, Metallurgy, and Exploration, Inc., 1997.

[30] Liang Y X, Fu H, Gu X Q. Provision of Occupational Health Services in China[J]. Asia Pacific Journal of Occupational Health &. Safety, 1998 (2): 1-5.

[31] Thomas L. The Handbook of Practical Coal Geology[M]. Chichester: John Wiley and Sons,2002.

[32] Kendell J. Natural Gas Annual 2006, Energy Information Administration [M]. Washington, DC : Office of Oil and Gas, USA Department of Energy,2007.

[33] Schatzel S J, Karacan C O, Krog R B, et al. Goodman G. V. R. Guidelines for the Prediction and Control of Methane Emissions on Longwalls[J]. Longwall Minging, 2008,50(50):632-639.

[34] Eghonghon O, Kelvin A, Megan O, et al. Coalbed Methane: Recovery &. Utilization in North Western San Juan[M]. Colorado: Department of Energy and Mineral Engineering in Penn State University, 2010.

[35] Du C Z, Liu W Q, Mao X B, et al. Drainage Methods of Coal Bed Methane and Its Application in Jincheng Mine Field in China[C]//Waste Management, Environmental Geotechnology and Global Sustainable Development International Conference. Ljubljana, 2007.

[36] Ozgen C, Karacan. Best Practice Guidance for Effective Methane Drainage and Use in Coal Mines United Nations Econornic Commission for Euroge Energy Series No. 31[J]. International Journal of Coal Geology, 2014.

[37] Cheng Y P, Yu Q X, Xia H C. Technical Methods of Safe High Efficient Exploitation of Coal and Gas Together in High-Gas Coal Seams in China [C]. International Workshop on Investment Opportunities in Coal Mine Methane Projects in China, 2003.

[38] Burton E, Friedmann J, Upadhye R. Best Practices in Underground Coal Gasification[R]. Lawrence Livermore National Laboratory, 2006.

[39] Chan M, Yano J, Garg S, et al. Coal Gasification[R]. San Diego Department of Mechanical and Aerospace Engineering Chemical Engineering Program, 2008.

[40] Lepinski J, Lee T, Tam S. Recent Advances in Direct Coal Liquefaction Technology[R]. Salt Lake City, Utah: ACS National Meeting, 2009.

[41] Williams R H, Larson E D. A Comparison of Direct and Indirect Liquefaction Technologies for Making Fluid Fuels from Coal[J]. Energy for Sustainable Development, 2003, VI(41).

[42] Karacan CO. Reconciling Longwall Gob Gas Reservoirs and Venthole Production Performances Using Multiple Rate Drawdown Well Test Analysis[J]. International Journal of Coal Geology, 2009, 80(3-4): 181-195.

[43] 钟似璇. 英语科技论文写作与发表[M]. 天津:天津大学出版社, 2004.

[44] 任胜利. 英语科技论文撰写与投稿[M]. 北京:科学出版社, 2006.

[45] 袁亮. 煤炭精准开采科学构想[J]. 煤炭学报, 2017, 42(1):1-7.

[46] Wang G F, Xu Y X, Ren H W, et al. Intelligent and Ecological Coal Mining as well as Clean Utilization Technology in China: Review and Prospects[J]. 矿业科学技术学报:英文版, 2019(2): 161-169.

[47] Zhou N, Jiang H Q, Zhang J X. Application of Solid Backfill Mining Techniques for Coal Mine Under Embankment Dam[J]. Transactions of the Institutions of Mining and Metallurgy: Section A, 2013, 122(4): 228-234.